Exploring the Big Salmon River

Quiet Lake to the Yukon River

Gus Karpes

ISBN 0-88839-422-5

Cataloging in Publication Data

Karpes, Gus, 1941 -
Exploring the Big Salmon River

Includes bibliographical references and index.
ISBN 0-88839-422-5

1. Big Salmon River (Yukon)—Guidebooks. 2. Boats and boating—Yukon Territory—Big Salmon River—Guidebooks. I. Title.
FC4045.B54A3 1998 917.19'1 C98-910165-7
F1095.B54K37 1998

Printed in Canada—Bari Graphics

Editor: Nancy Miller
Production: Ingrid Luters
Cover Photo: Gus Karpes

We acknowledge the financial support of the Government of Canada through the Book Publishing Industry Development Program for our publishing activities.

Published simultaneously in Canada and the United States by

HANCOCK HOUSE PUBLISHERS LTD.
19313 Zero Avenue, Surrey, B.C. V4P 1M7
(604) 538-1114 Fax (604) 538-2262

HANCOCK HOUSE PUBLISHERS
1431 Harrison Avenue, Blaine, WA 98230
(604) 538-1114 Fax (604) 538-2262
Web Site: www.hancockhouse.com *email:* sales@hancockhouse.com

Contents

Dedication

On the main street of Whitehorse, Yukon, there stands a lifesize statue of a prospector and his dog. The prospector is carrying a pack with the tools of his trade. This includes a rifle, a billy pot and a shovel. The dog is equally loaded down with a back pack. The statue was sculpted by Chuck Buchanan and was erected to honor the prospector. Among those on the honor roll plaque mounted on one side of the statue are such notable names as Edmund Bean, Jack McQuesten and Arthur Harper, all of whom were in the Yukon River Valley years before the Klondike gold rush. The dedication on the statue deals particularly with the nature of these early pioneers. It says:

"Dedicated to all those who follow their dream."

I would also like to dedicate this book to the early trailblazers and those who continue to follow in their footsteps.

Introduction

About 100 years ago the Yukon Territory was simply an empty area on most maps of North America marked unknown. By 1850, parts of the Yukon River and several of its tributaries had been surveyed and at least partially mapped by explorers of the Hudson's Bay Company, but by and large, to most of North America, the Yukon was still an unexplored hinterland. It was said to be a country like no other, and what little information was available described it as a distant, cold, unmapped, inaccessible and inhospitable territory somewhere above the sixtieth parallel near Alaska.

For a select few, this description in itself was a drawing card. Things were getting just a little bit too tame in the old west and the unknown north was just what the doctor ordered. McQuesten, Hess, Ladue, Carmack, Boswell, Schieffelin, Mayo and Harper are some of the names that come to mind when we speak of the early adventurers in the territory.

Oregon-born, Leroy Napoleon McQuesten came to the Yukon in 1871. He is perhaps best known for being the first president of the Yukon Order of Pioneers, the first real law and order society of miners in the north. Edward Lawrence Schieffelin was born in Pennsylvania. He came into the Yukon in 1881 and is best known for his discovery of Tombstone, Arizona. All of the mentioned men, and many still unnamed, had one thing in common. They were all in the Yukon Territory prior to the big gold rush of 1898.

It is hard for us to imagine what the Yukon was like when they arrived. It is difficult to visualize what manner of men they were, those who had the gumption to come into this unknown land. Whoever they were and whatever their reason for roaming about the north, many of us can identify with them. Step into your canoe, grab a paddle and push off into the current. Fish a mountain stream and cook the catch over an open fire. These are the age-old rituals that we can, as modern-day voyageurs, share with folks like McQuesten and his friends. As the night comes down around you and the loon's lonely call sends shivers down your spine, as the owl softly queries your presence around the camp fire, it is not hard to relate to these pioneers. They too heard the owl hoot its question into the night, they too heard the loon's haunting cry and they too wanted to be there.

Have a great trip.

Gus Karpes, 1997

1

A Short History of the Big Salmon River

The Big Salmon River is part of the Yukon River watershed. Much of its early history is directly associated with the exploration and discovery of the Upper Yukon River. For the sake of continuity and understanding, I am covering some of the Yukon River history in this guidebook.

The search for furs and minerals (mainly gold) was one of the driving forces behind the early exploration of the North American continent, and the Yukon Territory was no exception. In the early 1880s, the Hudson's Bay Company controlled most of the fur trade in Canada's North except for that part of the Yukon Territory described as the Upper Yukon River Valley. This area roughly encompassed the Yukon River drainage from its headwaters to Fort Yukon at the confluence of the Porcupine River, below present-day Dawson City.

In 1848, Robert Campbell of the Hudson's Bay established Fort Selkirk at the confluence of the Pelly and Yukon Rivers. In 1852, he and his crew were forcibly expelled by the Chilkat Indians who literally sacked and pillaged the place, to where the crew had no choice but to retreat. Some published accounts lead us to believe that this wanton destruction of the company's fort was carried out by hordes of painted and vengeful savages amid great bouts of gun fire in the commonly depicted tradition of the old west—this was not so. Campbell was already in poor condition as a result of having outdistanced his supply line. This had left him with few trade goods with which to barter for even the basic necessities in life.

Toward the end of the nineteenth century, the Upper Yukon River Valley was still very much an unexplored frontier. Despite Campbell's eviction, the Hudson's Bay Company coffers were receiving at least a part of the fur harvest of the area. The Chilkats, merchant Natives from the Alaska panhandle, made annual trading trips into the interior. On their return home, they traded the goods at tidewater with the crew of the *Beaver*, the company's steamer that made regular voyages along the Alaska coast.

The Chilkats came into the interior by climbing the coastal passes into the Yukon river system and used rafts to continue their voyage downstream. They returned to the coast in a long, heavy-laden trek through the same mountain passes. The most renowned of these routes was the Chilkoot Pass.

The interior trade was very lucrative for the Chilkats who had no intention of letting anyone in to contest the business with them. They forcefully and quite successfully defended their territory which remained under their merchant rule until 1880. At that time, they relaxed their vigilance or perhaps realized they could no longer control the traffic. In 1880, a group of miners recently out of the played-out goldfields of Cassiar, B.C., and led by spokesperson Edmund Bean, approached Chilkat chief Klotz-Kuch. Following a number of meetings, the two parties drew up a written agreement between them. This allowed Bean and eighteen other would-be miners access to the interior of the Yukon Territory. They were allowed to enter the interior via the Chilkoot Pass. In the agreement, among other things, the miners agreed, "to acquit ourselves as becomes orderly, sober, reasonable men," and not to carry "spirituous liquor...into the Indian country for the purposes of trade or barter with the natives." There is no doubt that there was a certain amount of gun-boat diplomacy involved in Klotz-Kuch's decision to relax the rules. The U.S. Navy vessel *Jamestown*, Captain L. A. Beardlee and his crew were on hand to assist in the negotiations and the drawing up of the agreement. Accompanying the miners to their meeting with Klotz-Kuch was Lieutenant E. P. McLellan and a number of the *Jamestown*'s officers. They were all under orders to wear uniform and side arms.

Although there are scattered records of individual miners and prospectors coming into the Yukon River Valley prior to 1880, it is generally recognized that this "negotiated" relaxing of coastal boundaries was the first significant and meaningful event in the opening up of the Upper Yukon River Valley.

In addition to Bean, there were eighteen others who signed the landmark agreement. That first year, all took part in an initial exploratory trip. As the season was almost spent by the time all of the outfitting was done, most did not get much beyond the upper reaches of the Yukon River. Several reached the Teslin River where they did some panning and prospecting, but all of them returned to the coast before winter set in.

Ironically, the celebrated exploration agreement created an industry for the Chilkats, probably more rewarding and financially lucrative than the long annual tramps into the interior. Used to packing heavy loads, the Chilkats were naturally hired to assist in the packing of goods from tidewater to the upper lakes. They, of course, charged a packing fee. This cash income continued well into the late 1890s with prices ranging from twelve cents per loaded mile in 1883 to as much as one dollar per pound, the charge during the height of the Klondike gold rush. Native youngsters and women were able to carry as much as seventy-five-pound packs, whereas adult males would take packs up to 150 pounds each.

Two of the miners that signed the 1880 agreement were George G. Langtry and Patrick McGlivichey. These two returned the following spring and this time reached the mouth of the Big Salmon River. Here they turned upstream for what they thought would be "about two hundred miles."

In the manner of prospectors, Langtry and his partner tested may of the river bars and tributaries for gold. They caimed to have panned "colors" on many of the bars of the Big Salmon River and a number of tributaries. Colors are traces of fine placer gold, likely deposited on the river bars at times of heavy water runoff such as one sees on most northern streams during the spring melting of ice and snow. The fine gold is carried along by the fast-running stream until the runoff is over or the river slows down. The heavy metal then settles on to the river bars. Colors therefore meant a source of gold upstream.

In that first season of prospecting, the two found traces of placer gold as they went but were unsuccessful in identifying the source of it.

The two could not resist naming their river and called it the Iyon, apparently naming it for a group of Natives that they found camped at the mouth of the river. (The local Tutchone name for the river is *Guy Cho Chs*, literally translated to mean *Guy* [salmon] *Cho* [big] *Chs* [water]. This does not in any way resemble Langtry's interpretation of the language. As Natives were prone to wandering around, perhaps those that Langtry and his partner met were just traveling through and spoke a different dialect or perhaps there was a total lack of perceptive monologue between the two parties.)

George Dawson of the Dominion Land Survey of Canada, allowed that this discovery of gold by Langtry and McGlivichey was perhaps the first recorded discovery of placer gold in the Yukon Territory.

Three years after this initial discovery, Lt. Frederick Schwatka (U.S. Army), made his celebrated raft trip down the length of the Yukon River. According to Schwatka, his Chilkat guides call it the *Tah Heen A.* Schwatka felt that this name too closely resembled the name that they had given to the river close to present-day Whitehorse, the Takhini River. He renamed the river the d'Abbadie, after French explorer Anton d'Abbadie.

In 1887, George Dawson, DLS, the leader of the Yukon Expedition and the man after whom today's Dawson City is named, sorted out the naming process and officially adopted the name Big Salmon River. The names Iyon and d'Abbadie disappeared into history, although d'Abbadie was later given to a mountain on the North Fork, a tributary of the Big Salmon River.

While Dawson's group was camping at the confluence of the Yukon and Big Salmon rivers in 1887, they met a party of miners. The group was led by a New Brunswick prospector named John McCormack. McCormack and his outfit had spent the entire summer of 1887 on the Big Salmon River. They had pushed, poled and portaged upstream as far as Quiet Lake. He and his crew were responsible for many of the geographical names along the Salmon route.

Unfortunately, we don't know much else about McCormack and his crew. There appears to be little doubt that they were a salty lot of individuals. They literally traveled into the unknown to explore this tributary of the Yukon River without any inkling as to what was at the other end of their selected waterway. No doubt they had their own reason for picking the Big Salmon as their goal or perhaps they had the 1881 information of Langtry and his partner.

McCormack and his crew where much better navigators and judges of distance. Their description of the river, the lakes and the distances involved, came uncannily close to the actual measured distances and descriptions of the Canadian Geographical Survey parties who explored the area in later years.

Until the early 1940s, access to the Big Salmon River was via the Yukon River. With the building of the Alaska Highway and the Canol

The Big Salmon River. *Photo: Scott McDougall.*

Road, passage to the Big Salmon River was available from its headwaters at Quiet Lake. Since that time, the river has seen a steady stream of river travelers.

Mineral exploration along the river and its tributaries such as the South Fork and North Fork has been ongoing since Langtry and McCormack's first prospecting trips. A substantial discovery at Livingstone Creek, a tributary of the South Fork, created great furor during the late 1800s. This had little effect on the main river, as the main supply route to Livingstone was from Mason's Landing on the Teslin River from where an overland trail led to the gold town of Livingstone Creek.

Today, an adventure trip on the Big Salmon River is still an exhilarating and challenging experience. Here and there you will encounter signs of civilization, but for the most part the wilderness still prevails. You can still savor the same thrill of it all and feel the same sense of accomplishment that McCormack and his companions felt more than a century ago.

2

Traveling Tips

Access and Takeout

The South Canol Road provides access to Quiet Lake, one of the lakes that make up the headwaters of the Big Salmon River. Canol Road leaves the Alaska Highway at the riverside community of Johnson's Crossing, eighty-five miles (136 km) south of Whitehorse. Start your trip from the public campground about six miles (10 km) from the north end of Quiet Lake. For those who are unfamiliar with Canol Road, it is the second campsite along the shore of the lake. The South Canol is a beautiful drive but can get very tricky in adverse weather. The better part of a day should be allowed for the drive from Whitehorse.

The conventional takeout is at the Village of Carmacks at mile 200 (km 320) of the Yukon River. You can continue on the Yukon River to Dawson City, a further 260 miles (416 km) downstream.

The Big Salmon River. *Photo: Scott McDougall.*

Distances

By road: Whitehorse to Quiet Lake – 155 miles/248 km
Carmacks to Whitehorse – 110 miles/176 km

By river: Big Salmon River inclusive of lakes – 147 miles/235 km
Yukon River from Big Salmon Village to Carmacks
– 75 miles/120 km

You should be equipped to handle your own emergencies. *Photo: Scott McDougall.*

Remoteness

You are on your own once you leave the Quiet Lake Campground, and if you are traveling early or late in the season, count on little company. Once you are through the lakes and committed to the Big Salmon River proper, there is no turning back. You should be equipped to handle your own emergencies.

The Big Salmon River is not considered to be an amateur river, especially during the early part of the summer when it is in flood. The upper reaches are fraught with fast water, sweepers, log jams, rocks and rapids. I am not trying to discourage you, but do suggest that you have some experience prior to starting out or alternatively consider traveling as part of a small group.

Should an emergency occur which prohibits you from carrying on, stay put! Under no circumstances try to hike out! Wait until someone else comes along. Carry a small emergency pack on you at all times. I suggest using the small waist or fanny packs which can be worn at all times with little discomfort. Whatever else you carry in this pack, make sure it contains a generous amount of mosquito repellent and several disposable lighters.

Report to someone before you depart and give them an anticipated return date. Make sure that you check in with them when you finish the trip.

Water Levels

Quiet Lake is generally free of ice in late May or very early June. Some years this break up has been delayed by as much as two weeks into the middle of June due to an unusually cold winter and spring. If there were a record snowfall and an early hot spring, high water levels will prevail during most of June. Canoeing the river during this period is not recommended unless you are an experienced white water canoeist.

Drinking Water

Recently the affliction known as "giardiasis" and commonly referred to as beaver fever has become of concern. The *Giardia lamblia* parasite, the cause of it all, is found worldwide and is one of the most commonly reported human intestinal parasites. Although it can be transmitted on food and from person to person, it is frequently transmitted through surface water. And though the parasite is carried by all mammals, the beaver seems particularly susceptible to it, hence the name beaver fever.

The symptoms of giardiasis are not pleasant. Frequent stomach cramps, diarrhea and a general queasy feeling come and go. As the parasite has an incubation period of ten days to two weeks, the problem may not come to light until you are home. The best cure, of course, is to avoid getting it in the first place.

Keep clean! There is no reason to drop normal sanitary habits during a wilderness trip. Carry a tube of biodegradable soap with you at all times. Consider treating your drinking water or carrying a small water filter and pump with you. There are many of these small gadgets on the market. Tincture of iodine, available at most drug stores and chlorine are still two of the most dependable treatment products available to you. Modern household bleaches no longer have chlorox as part of their makeup and can no longer be used.

Garbage

If you can carry it in, you can carry it out. Contrary to popular opinion, aluminum foil does not burn. Burying your garbage or disposing of it by throwing it in the river is totally unacceptable behavior. Burn paper products and leftover foods. Scorch foils and cans to deodorize them before putting them in your garbage container.

Pack out a little more than you brought in. If we all follow this practice, the enjoyment of the wilderness can be appreciated for many generations to come.

Camping Etiquette

"Where can we camp?" This is not an uncommon question from those who are taking a wilderness river trip in the Yukon for the first time. It is one of the last places in North America where you can still camp where you wish, and many of our national parks and recreation areas allow camping in designated areas only.

There are no organized or specified campsites along the Big Salmon River. The picturesque setting of a site, its ease of access and its strategic location will sometimes make it appear as though it has been designated as a camp, and you will find some conveniences already in place. Unless you are invited to share a campsite by those already occupying the spot, don't crowd in.

Do not indiscriminately cut and hack away at trees and other foliage that surround a prospective site. Normally there are plenty of driftwood and dead falls to supply your needs. Think about the length of time that it takes for a healthy tree or bush to grow in the harsh climate of the North and you will naturally be a little more careful with this resource.

Take extreme care with the placing of a campfire. The haphazard, casual treatment of this privilege has been responsible for the destruction of acres of forest. Place your campfire in such a way that an underground fire has no chance of starting. An old fire pit does not necessarily qualify as a safe place as you may simply be compounding the risk. Make sure that the fire is completely out before you leave a site. There is lots of water out there!

If you do end up sharing a campsite, do not build a second fire. If you are sharing the camp, share the fire. There is no reason to disfigure a campsite by creating rock and ash piles all over the place.

Bring a folding shovel or gardening trowel with you. Bury human excrement and burn all toilet paper. If you're squeamish about bringing the paper back to camp, use a paper bag to carry it in.

Hunting and Fishing

The lake system and the Big Salmon River have some great fishing opportunities. A Yukon fishing license is required. Lake trout and the Arctic grayling are plentiful in the upper lake system and the connector streams. The grayling, northern pike and several species of white fish, or Inconnu, can be caught in the river itself.

Fishing regulations change annually and in many lakes and streams of the Yukon, barbless single hooks are compulsory.

There is an annual run of chinook salmon into the Big Salmon River. The run starts in late July and can go until the end of August. These fish have traveled almost 2,000 miles (3,200 km) to get here. As soon as the salmon hit fresh water, they start to deteriorate and rarely feed, totally intent on that spawning run. The salmon in the Big Salmon will have turned red and are not at all what you envision as an eating delicacy. Although it is legal to catch these salmon, I feel that there is much better fish out there for eating. They should be left alone and allowed to procreate in peace.

If you are taking a trip in the fall and the water has receded, you can clearly see the salmon eggs on many of the exposed gravel bars of the main river. Be careful where you walk as the eggs will be destroyed if stepped upon.

If you wish to carry a firearm for self protection, you may carry a "long gun" (any legal firearm except for a handgun) on your trip. Hunting is prohibited except for small game and then only in season. A small-game hunting license is required.

Bear repellent, or bear spray as it is commonly referred to, is available in stores and from canoe outfitters in the Yukon. I have seen it tested and it works to discourage any animal. It does require close quarters to be effective. Your best defense is always to avoid the encounter totally and use common sense in the choosing of a campsite and in the handling of all foods and particularly fish.

It is totally unacceptable and illegal to hunt large game in the Yukon, unless you are a Yukon resident or are accompanied by a registered hunting outfitter, and then it is allowed only in the hunting season. The season generally runs from August 1 to September 30. All regulations change from year to year and it is wise to check what the regulations are before leaving home.

Inquiries can be sent to: Government of the Yukon, Dept. of Renewable Resources, P.O. Box 2703, Whitehorse, Yukon, Canada Y1A 2C6. The Yukon government has an active patrol and enforcement branch.

Weather

Generally speaking, the Yukon is considered to have a semiarid climate. But many times I have wondered who gave it this classification. There are several appropriate sayings that we Yukoners have when talking about the weather.

"If you don't like the weather, wait for fifteen minutes and it will change for you."

"Any fool can be uncomfortable."

During the summer months, you can expect up to twenty hours of daylight per day. It never really gets dark during the months of June and July. After having to deal with an equal amount of total darkness during the winter months we all, including the animal populations, go into a summer mode. This means that you stay awake at all hours and cram as much living as possible into the summer months. We have midnight golf tournaments, baseball games, fishing derbies and what have you. Outsiders have a little trouble with the perpetual daylight, but we love it.

The climate around the world has been changing. As I am writing this, it is well into December and the thermometer has yet to drop to minus twenty Celsius. There is very little snow and Lake Laberge, where I live, is wide open.

Expect to deal with fast changes of weather daily with a shower or two mixed in. Bring rain gear. Do not be cheap in choosing it as you will be in and out of it several times a day. Take every opportunity to dry out your camp. Even under ideal conditions, a sleeping bag will collect moisture. Carry a bag that is rated at a minimum of zero degrees.

Bring several hats, long-sleeved shirts and a coat. I have been caught out in a snow storm in June and late August. I have also been out in eighty-degree weather as early as May.

Equipment

There is so much new equipment on the market today that I won't even try to recommend or produce a list for you. I bring a camp stove. I have a variety of them and the type and length of trip dictates which one I use. A camp stove will work in the rain, you don't need dry wood to start it, and in the morning it makes coffee pronto.

I also pack a supply of fire starter. Again, this comes in a variety of packaging. I try to stay away from the petroleum-based products as they tend to contaminate whatever else you carry in the same bag. Disposable lighters are great. I have been underwater with one many times. A little shaking and blowing and they have always worked.

Do not leave without mosquito repellent!

A hatchet and/or Swede saw, sometimes referred to as a bow saw, save a lot of wear and tear on clothing, knees and feet. If you want to leave one of them at home, leave the hatchet.

Bring a variety of plastic bags. Buy a good quality. The large garden ones will do for an emergency rain coat. The smaller ones for garbage, laundry and waterproofing.

I use a net bag to keep fish in until I am ready to cook it. I also use the bag to cool off foodstuffs. Make sure the bag is anchored when dunking it in the current.

The rest is optional and up to you. It depends largely on your personal preferences. I never count on catching fish in my menu. I always pack an extra emergency meal. All of my drawstrings have cord locks. All of my containers, coolers, stoves, etc., are secured with a good-quality strap.

Bears

I have included this specific heading in every book as it is by far the most common subject brought up for discussion by all who travel the river. Please read it carefully and take the proper precautions.

In the more than thirty years that I have spent on the rivers, I have never had a problem or a serious incident involving bears. This does not mean that I have not seen them during my travels and it does not mean that there have been no incidents.

Bears have lived in the Yukon for thousands of years. We can avoid, or least minimize, the chances of an encounter and/or injury by understanding some of their characteristics and their lifestyle.

The most common sighting or encounter is with the black bear (*Ursus americanis*). There are an estimated 10,000 of these animals throughout the territory. They are concentrated mostly in the central and southern Yukon. Although we refer to this species as "black," it can range in color from pure black through various shades of brown to cinnamon or blonde. The coloring is more consistent on the black bear than the multicolor common to the grizzly bear. The male black will reach an adult weight of about 110 kg (about 250 lb.), while the female at maturity will weigh about 75 kg (170 lb.).

The grizzly bear (*Ursus arctos*) ranges in color from dark, rich brown to almost blonde. The coloring is not consistent, with the legs generally being darker. The longer body hairs, particularly those around the mane, have lighter tips, hence the name "silvertip" or grizzly for its consequent grizzly appearance. There are an estimated 6,000 to 7,000 grizzlies spread throughout the entire Yukon Territory. It is a considerably larger bear than the black. The average adult male can reach about 200 kg (450 lb.), a female reaches about half that weight.

Bears will eat almost anything. Although a typical diet depends on the bear's territory, the menu can consist of roots, berries, grubs, small rodents, eggs, carrion and, of course, fish.

Generally speaking, all bears tend to range at the higher elevations. This is particularly true for grizzlies which prefer the subalpine areas. The river trails are visited during the early spring and during the salmon spawning season later on in the summer. As the snow line recedes, the bears tend to range higher in the more prolific growth areas where heavy bush does not hinder the growth of their primary food sources.

Bears will always investigate new food sources and have an acute sense of smell and hearing. Contrary to popular belief, a bear has good eyesight but tends to identify by its better senses of smell and hearing, thereby creating the impression that its sight is poor.

A "spoiled" bear is one that has learned to identify humans as a food source. Generally speaking, all bears sighted or encountered during a wilderness river trip have not made this association and are considered wilderness bears.

Following are some suggestions for campers traveling through bear country:

1. When carrying fresh foods such as fruits, meats, vegetables or sweets, carry them in an airtight container.
2. Cook enough food for one meal only, and if there are leftovers do not leave them out. If you cannot put them in an airtight container, burn them immediately.
3. Scorch all noncombustible materials such as cans and tinfoil before putting them in your garbage container.
4. Meticulously clean up after every meal. Use a little bleach or chlorox in your dishwater.
5. Do not store foods in your tent.
6. Clean fish away from camp if possible. If cleaning must occur in camp, make sure that all traces of the innards and leftovers are cleaned up and incinerated.
7. Never bury garbage, pack out everything that you brought in. If you were able to pack it in, there should be more than enough room to carry it out!
8. Personal cleanliness is also important. A pair of pants you've worn, wiped your hands and knife on, gets to smell quite good to a bear on day ten. Bring a plastic airtight laundry bag for soiled clothing.

A word of caution to women. Try to stay out of bear country during your menstrual cycle. There is some evidence that bears are attracted to women during this physiological condition. If this is unavoidable, keep scrupulously clean. Acquaint your partner or fellow campers with your problem so that you will have the time and privacy to deal with it. Incinerate used sanitary napkins or tampons as soon as possible. Carry a supply of ziplock plastic baggies and some paper lunch bags with you. These will do as temporary receptacles en route until a fire is available.

Pay attention to your surroundings. Patrol a campsite prior to setting up camp. If the spot is an obvious feeding area for a bear, continue on to the next suitable site. You can smell a bear in the woods. Their excrement and body odor is very pungent, particularly in late summer.

Bear Encounters

Bears, like dogs, all have their own personality and there is no one answer to the handling of a bear encounter. Generally speaking, a bear will avoid an encounter with you if it knows you are there and you are not competing with it for food. Let the bear know you are there.

A bear standing up, slapping its sides and waving its nose around is trying to identify you. Keep facing it, and do not turn your back and run. Talk to it and announce your presence, slowly retreating and giving it room. Do not trap a bear, and by this I mean always give it an avenue of escape. Do not contest the right to food even if you are the one that bought it at the supermarket.

Never attempt to get close to a bear cub. Although mama may not be visible, she is around and will do anything to prevent harm coming to her cubs and you present a threat to them.

I have found that all the other animals in the bush can give us advance warning. Chattering squirrels, ravens, jays, etc., can herald the approach of a bear or other large mammal such as a moose. Prior to the large animal entering camp, all the normal noises seem to reach a crescendo, followed by an almost eerie period of total silence before the large animal announces itself.

To conclude, I do not wish to create the impression that bear problems are inherent to and an everyday hazard of a wilderness canoe trip. Bear problems are rare. Sightings are not a daily occurrence. With these comments I hope some of the mystery has been dispelled. Bears are quite willing to share the wilderness with you, as long as you recognize its right to be there. Sightings of bears and other wild animals are something to be enjoyed as a rare bonus to the trip.

3

The Major River Sections

The Canol Road

MAP 1, 2 **PAGE 39, 40**

Part of the Big Salmon River experience is the drive to Quiet Lake on the South Canol Road. It is a beautiful outing even on a bad day.

Like the Alaska Highway, the Canol Road was built by the U.S. and Canadian Armed Services during World War II. The road was built as a service and construction road for a pipeline which connected the oil fields of Norman Wells, N.W.T, to an oil refinery in Whitehorse, Yukon. The finished petroleum products were then moved to Alaska via a pipeline along the Alaska Highway corridor. The name CANOL stands for Canadian Oil.

The South Canol Road runs from Johnson's Crossing on the Alaska Highway to Ross River, Yukon. The North Canol Road runs from Ross River to Norman Wells, N.W.T. The road and pipeline were completed in 1944. The Pacific war theater was then starting to go in favor of the U.S. and as a result the pipeline saw very little use. It was dismantled and for a time, the South Canol Road was maintained into Ross River. It was the only road access to that community until the mid-1960s.

In the late 1960s, a large lead/zinc mine was built at Faro, a few miles from Ross River. At the same time, the new Campbell Highway was built which connects both Ross River and Faro to the Yukon communities of Carmacks and Watson Lake. The South Canol Road is now open during the summer months only.

Quiet Lake to the Big Salmon River —Mile 0 to Mile 12 (km 0–19)

MAP 2 **PAGE 40**

The exit out of Quiet Lake is at the northeast end. When paddling the lake, there is no need to be too far offshore at any time.

Generally speaking, Quiet Lake lives up to its name. Storms, when they occur, don't last too long. The east shoreline does not have many places to land in an emergency. If there is a high wind blowing, don't start out until it has subsided.

If you have a late start out of Whitehorse, it is best to stay at the lakeshore campground for the first night and get an early start in the morning. There are those who proceed posthaste to the end of Quiet Lake before settling in for the night. I have an aversion for hurrying on a trip such as this and generally spend the first leisurely evening around a campfire or perhaps fishing the lake for trout or grayling. I then get a relaxed start in the morning which sets the pace for the rest of the trip.

There are few opportunities for camping on the east shore of the lake. I have marked the exception. There are a number of good camps at the end of the lake but nothing along the stream connecting Quiet Lake to Sandy Lake.

Sandy Lake is generally quiet, well behaved and an easy paddle. The stream connecting Sandy to Big Salmon Lake is comparable to the previous stream but again offers little in the way of a campsite. If you missed catching your dinner in the first stream, you get a second chance here for both grayling and trout.

Big Salmon Lake can have some fierce winds blowing, and if the weather looks at all menacing, enter the open lake with caution. There is a campsite immediately to your left upon entering Big Salmon Lake.

Power boats should have no difficulty in going down as far as Big Salmon Lake but I have always had to push and pull my way back up in the stream connecting Quiet to Sandy Lake.

Quiet Lake

MAP 2 **PAGE 40**

Quiet Lake was named by John McCormack and his crew during their trip in 1887. Quiet, I assume, is the way they found the lake during their stay and it would have made quite an impression on them after their upstream voyage. The lake is nineteen miles long and about two miles wide at its broadest point.

It is only since 1944, when the Canol Road was completed, that we have had easy access to Quiet Lake and the headwaters of the Big Salmon River. Until then, the only way to get to the lake was upstream from the Yukon River.

If you are camped on the lake shore for the first evening, a continuous mechanical thumping noise is heard at all hours of the day and night. Look up the hill directly east of the campground. The Yukon Department of Highways has a maintenance building and equipment yard on the hill. The noise is a large diesel generator which runs continually during the summer months that the road is open.

Erik Nielsen, the brother of actor Leslie Nielsen of the *Naked Gun* series, was for many years the member of parliament for the Yukon Territory; he has a retirement home at the south end of Quiet Lake. There is no access and quite often you will see a floatplane fly into the lake. In all likelihood it is headed for Mr. Nielsen's cabin.

Sandy Lake

MAP 2 **PAGE 40**

Sandy Lake, I assume, was also named by McCormack and friends. It is about two miles long. It is quiet and surrounded by some picturesque mountains. There is a campsite on the promontory about three-quarters of the way through the lake. The quiet, still water of the slough behind the campsite is the home of some ferocious northern pike.

Sandy Lake. *Photo: Scott McDougall.*

Brown Creek

MAP 2 **PAGE 40**

This flows into Sandy Lake at its southwest end. Old records denote that this was named "for a man named Brown who mined the creek."

R. G. McConnell, in a report to the Geological Survey of Canada writes: "He (Brown), is said to have made sufficient off the property for several grubstakes, mining primarily the head of the canyon just off the lake shore."

Perhaps now is the time to try your gold pan at Brown's old workings.

Old election lists show an A. N. Brown, as a resident of the Hootalinqua-Livingstone riding during the early 1900s. This area fell into that census district, and as he is the only Brown listed between 1898 and 1916, we can assume that this was the man in question. We have no further information.

Big Salmon Lake

MAP 2 **PAGE 40**

Big Salmon Lake is about five and a half miles long. Your entry into the lake is halfway along its south shore.

This was originally called Island Lake by the McCormack party for the large island in the middle of the lake. It was renamed Big Salmon Lake by members of the Geological Survey crew. The same survey crew for some strange reason decided to measure the depth of the lake and recorded it as 138 ft. (43 m) deep. I don't know at what point in the lake they measured the depth.

Wood is generally scarce at the first campsite into the lake. If weather permits, choose an alternate site, perhaps across the lake.

Nisutlin Bay Outfitters of Teslin, have a hunting camp about halfway along the south shore. It is a pleasant place to camp. Respect private property, watch your fire and clean up before you leave.

Big Salmon Lake is on the flight path of migrating geese and swans. One late September evening in the 1970s, I was camping at the lake and found myself virtually surrounded by the large birds. They kept up a constant chatter all night long which was like going to sleep with the radio on.

Tower Peak—5,672 feet (1,730 m)

MAP 2 **PAGE 40**

This peak is on your right, on the north shore as you enter Big Salmon Lake. Arthur St. Cyr of the Dominion Land Survey named the peak in 1901.

Mount St. Cyr—6,725 feet (2,050 m)

MAP 2 **PAGE 40**

The tallest peak is the one that St. Cyr named after himself. Arthur and his crew mapped the area for the Canadian Government in 1901.

Among St. Cyr's other accomplishments was the surveying of the overland Stikine–Teslin route to the gold fields of the Klondike and the surveying of the British Columbia–Yukon border.

Caribou Creek

MAP 3 **PAGE 41**

This comes into the Big Salmon from the north out of the valley to the left of Mount St. Cyr. The creek is about sixteen miles long. McCormack and his party named the creek. It enters the river through a low-lying marsh at the outlet of the lake.

The Big Salmon River

MAP 3 **PAGE 41**

Congratulations, you have reached your goal. At this point, it would be wise to ensure that you have no loose gear in the canoe. Make sure all your important things are in waterproof containers and everything is loaded properly for the trip ahead.

Moose Creek. *Photo: Yukon Government photo.*

Big Salmon Lake to Moose Creek —Mile 12 to Mile 45 (km 19–72)

MAP 3, 4, 5 **PAGE 41, 42, 43**

Immediately after you enter the river proper, a cabin appears on the left bank. In bad weather it is a great place to dry out.

Once past the cabin, you are encountering what I feel is the most demanding part of the trip. This is partly because you are getting your first taste of some very arduous paddling while your canoe is still heavily loaded, but also because the river is at its narrowest with many tight corners. Characteristically the river ahead twists, turns and is full of slumping banks, sweepers sticking into the river and log jams. Sweepers, trees that topple into the water, will continue to be a problem for the first several days. Take your time.

A jam almost completely blocks the river about two miles down. This will require a short portage during lower water levels. In 1995–96 someone took a chain saw to some of the logs in the jam which opened it up somewhat. I have heard the comment that it is now worse as you do not get enough warning or time to line up for the passage that the person attempted to create through the jam. Again, take your time and stop to have a look before proceeding.

Once through the jam, the general features of the river remain the same and do not noticeably change for the first few days. The disorderly and unannounced timber sticking into the river stay with you, although the bends start to open up a bit and the river gets a little wider with the addition of the water that the creeks are bringing in.

Camps are hard to come by and the narrow confines of the river valley do not lend itself to an early sunrise. Clay banks and rocky bluffs will start to appear for the first time on day two. A short rocky stretch of river gives you a taste of what is to come.

Sheep Creek and Moose Creek can be a little difficult to identify as a great number of streams and rivulets spill themselves into the main river. Sheep Creek is normally a short, intermittent tributary four to five miles long. During runoff periods, it can be a fair-sized, swollen and quickly running stream. Moose Creek originates out of Pleasant Lake, twelve miles north. It enters the Big Salmon in a Y.

Gray Creek/Scurvy Creek—Mile 17 (km 27)

MAP 3 **PAGE 41**

These enter the Big Salmon as one. The split between the two is approximately four miles upstream. Both creeks are about twenty miles long and have their source in the mountains of the Big Salmon Range to the west. If you have been encountering rainy days and stormy weather, the creek will be swollen and driftwood may pose an additional hazard.

I have no official knowledge about the naming of Gray Creek. Again, referring to the voters list for the early 1900s, I find an Edward W. Gray listed as a miner in the district from 1909 to 1916. I assume that this creek was named for him.

Scurvy was a frequent disease caused by an inadequate diet or more specifically the lack of ascorbic acid in the rations (vitamin C). Scurvy Creek is a common name in the North and generally the name implies that one or more people contracted the disease while living there.

R. G. McConnell in his report to the Geological Survey of Canada in 1898 reports: "Fine colors were obtained in the wash of many streams emptying into the Salmon, and on a bar at the mouth of the stream that joins the Big Salmon three or four miles below the lakes, a very good prospect was obtained."

This find of McConnell's and his consequent report, which would have been available to the public, were enough to send some on their way into the creek in search of more of the same. It can be assumed that at the time of McConnell's report the creek was still unnamed.

Moose Creek—Mile 45 (km 72)

MAP 5 **PAGE 43**

Moose Creek is another one that was named by McCormack. It is most likely that they spotted a moose in the vicinity which gave rise to the name.

Moose Creek to Souch Creek —Mile 45 to Mile 72 (km 72–115)

MAP 5, 6, 7 **PAGE 43, 44, 45**

Short, fast runs of rapids are the order of the day. Sweepers are still present but more easily avoided as the river widens and the curves open up even more. The river bottom starts to change to rock and gravel with an occasional boulder thrown in for good measure. Beware of the ever-present log jams. As the water flows through these heaps of driftwood, you do not get the benefit of a deflecting current, and they become a real trap.

Now it must be evident to you why I do not recommend the Big Salmon River in the flood season.

Some previous experience allows you to have at least half a chance to make it through to the bottom end unscathed. Traveling with a group can make up for the inexperience and also ensures that you always have someone who can render assistance.

This is not an idle warning. Let me relate one of my more memorable experiences on the river during the years that I was an outdoor guide in the Yukon.

An old friend and companion on many river excursions, became a little piqued with the uneventful and somewhat docile canoe and boat trips she had been taking with us. Despite my discouraging her and my emphatic "Nos" when she asked for a trip on the Big Salmon, in a moment of weakness I agreed to the venture. Not only did I agree, but compounded the mistake by agreeing to an early June timetable and made things worse yet by making it a solo trip—one canoe.

My companion was in her early sixties, physically fit but plagued with the normal afflictions that age endows us with. Your knees don't work like they used to, the right shoulder twinges under strain, your hearing isn't what it used to be and the canoe seat gets harder and harder as the trip progresses.

The Salmon was high and in rushing flood. The weather was deplorable; it rained the first three days of the trip. Frequently we had to share our somewhat limited channel with two or three large floating logs which were shooting down the river just as fast as we were.

On the third day, coming around a tight S-curve in the river, a large, drifting log completely blocked our path. By the time the situation was analyzed and orders passed ahead to my slightly deaf companion, a tangle with the log jam was inevitable. I did manage to launch myself onto the pile of logs and lever my companion out of the clutch of the current, just before the canoe disappeared into the depths. Instinctively I had also managed to hold onto my paddle and grab the spare on the way out, although I don't know what I was going to be using them for at that point.

Once the adrenaline quit pumping, an assessment of the situation was not too encouraging. If you can imagine two wet and bedraggled-looking individuals, sitting on a pile of firewood, completely surrounded by a torrent of water, with little more than our emergency packs and two paddles, you can understand that we were definitely unimpressed with our predicament.

Both being smokers and having the foresight to carry this most important commodity on our person, there was nothing to do but sit and evaluate the situation enveloped in calming clouds of nicotine. Only fellow smokers can possibly relate to the comforting feeling of this activity amid chaos.

The canoe was trapped some six feet down, pinned against the log jam. It had not broken up, as the large waterproof orange packs were still jammed inside and supported it structurally.

Now I swear this is a true story. After my calming smoke, I nonchalantly grabbed one of the smaller trees off the pile, fished out the bowline, and tied it to a log anchor of which there were plenty. I took the pole, reached into the river and levered on the canoe. Magic! The canoe shot up out of the depth, hit the end of its rope, spilled the packs into the river, did several graceful pirouettes in midair and landed in the water at my feet, dry and right side up! I tried to hide by total look of disbelief at this fortunate turn of events. Although it must have been obvious to my companion that this overwhelming success at fishing was out of the ordinary, she had the good grace to do nothing more than congratulate me with a believable, "Oh marvelous!"

The canoe, slightly worse for wear, appeared watertight and we were able to continue our perilous journey minus some of our top-loaded items. I picked up our waterproof packs some distance downstream and cooked in a coffee can for the rest of the trip, but came spurting out of the bottom end in good form.

My friend and I have some fond memories of our Salmon incident. There is little doubt that we were lucky and the situation could have been much worse.

So you see, having experienced such an incident, it is with some authority that I pass on words of caution to you.

Souch Creek—Mile 72 (km 115)

MAP 7 **PAGE 45**

Souch Creek is about the same size as Moose and Bat creeks, and it comes out of the high peaks to the north. It is named for Flying Officer George Allan Souch, DFC, RCAF. He was born in Toronto, Ontario, in 1916. He served in World War II and died overseas on July 29th, 1943. I am not aware of what his connection was to the creek or the Yukon.

Souch Creek to South Fork —Mile 72 to Mile 109 (km 115–174)

MAP 7, 8, 9, 10, 11 **PAGE 45, 46, 47, 48, 49**

It now appears as if someone took a handful of boulders and scattered them indiscriminately along the river bottom. At high water most of them can be traversed but at the lower water levels you will have to wend your way between them.

As you emerge from the confines of the narrow river valley which has dominated the scenery to this point, the rolling Seminof Hills come into view. Camping opportunities become more plentiful. If you have had some bad weather in the upper reaches during the past few days, things should improve. Coming out of the valley and into the wider proportions almost seems to be synonymous with a change in the weather. It also seems, to me anyway, that you can smell the South Fork approaching—the odor emitting from the marshes and swamps bordering the river. Just before the South Fork you will notice a road appearing on both banks. This is the crossing of a recently constructed winter road into Livingstone Creek about twenty miles upstream on the South Fork.

There are great camps on both sides of the Salmon at the South Fork confluence. If you plan on landing on the left (west) shore, watch for a large whirlpool just past the confluence. Stay out of it as it is strong enough to capsize a canoe.

The South Fork—Mile 109 (km 174)

MAP 11 **PAGE 49**

At the turn of the century, placer gold was discovered at Livingstone Creek, Little Violet Creek, St. Germain and Summit creeks. These are all tributaries of the South Fork and centered about twenty miles upstream. The discovery came at a time when all eyes and interests were focused on the Klondike and the Livingstone area find did not receive the worldwide publicity that it deserved.

Since its discovery, the region has been mined continually albeit on a sporadic basis. It was estimated that in the year 1902, all of the creeks produced between $30,000 and $50,000 in gold. This was a very substantial amount of cash in those days when the price of gold was around $30 per ounce.

The normal route into the Livingstone area was, and is, via the Teslin River. Mason's Landing, about twenty miles upstream, was considered to be the head of navigation on the Teslin River for the larger sternwheelers. An overland summer trail led from there to the South Fork of the Big Salmon River. The Big Salmon River was rarely used as a transport corridor for Livingstone traffic.

The Seminof Hills

MAP 11, 12, 13 PAGE 49, 50, 51

Once at the South Fork, a range of distant hills comes into view. Seen from a distance, I always think they appear somewhat purplish in color, which of course is an illusion as they are quite green when you approach them more closely. The Yukon River Valley is on the other side of this range of hills. Lt. Frederick Schwatka (U.S. Army) named the hills in 1883 during his raft trip down the length of the Yukon River. In Schwatka's book printed in 1894, he spells the name Semenow. I have also seen it spelled Seminov. In any event, the hills are named for Von Seminow, past president of the Imperial Geographical Society of Russia.

South Fork to North Fork —Mile 109 to Mile 135 (km 174–216)

MAP 11, 12, 13 PAGE 49, 50, 51

As you pass the South Fork, the river broadens and assumes a more leisurely pace. The distant low profile of the Seminof Hills makes for an early sunrise and a longer evening.

Occasionally there is a fast run of water as the river chooses its channel between gravel bars and occasional islands. High clay banks now border the river.

As you approach the North Fork, the river bottom again changes. Rocks and boulders foretell the approach of more rapids. The rapids are long and full of obstructions but should not really present a problem. The runs are straight, and the large boulders can be seen in plenty of time to be avoided. At high water this can be quite a ride. On one such trip, I can recall the North Fork, also swollen, completely running over its banks and spilling over the banks of the main Salmon River to where it was a little difficult to choose a likely route in the time allowed.

There are a number of good open campsites near the North Fork on both sides of the river. Teraktu Creek offers little room for a camp unless it is at low water. At other times I recommend you pass it up as an overnight stay.

Teraktu Creek—Mile 117 (km 187)

MAP 11 **PAGE 49**

This is one of the few creeks and places that carries an original, untranslated Indian name. The original Tutchone name means "sharp rocks sticking out." Apparently this referred to the mountains from which the creek flows. The creek is about twenty-five miles long and flows in from the mountains to the east.

One of these mountains is Mount D'Abbadie, the name that Schwatka tried to attach to the Big Salmon River itself.

The North Fork—Mile 135 (km 216)

MAP 13 **PAGE 51**

In the geographical accounts of 1887, it was reported that a large Indian encampment was located on the upstream or west side of the North Fork. This was apparently about half-mile upstream and a great number of salmon had been speared by the Natives who were smoking them on large racks built along the river.

Into the late 1940s, the site remained as a favorite fishing and hunting camp for the Natives that lived at Big Salmon Village below. There was an overland trail between the North Fork and the village on the Yukon River.

North Fork to the Yukon River —Mile 135 to Mile 147 (km 216–235)

MAP 13, 14, 15 **PAGE 51, 52, 53**

Rapids are still the order of the day. Handling them seems to be old hat after the turbulent run to the North Fork. The stretch of river just past the North Fork shows an easily visible drop in elevation.

During the last few years, the river has changed in the area just past Illusion Creek. The exact changes do not appear on the map. It seems that one of the curves got jammed with ice or debris and the river dug out a short cut. This could again change if the obstruction clears itself in a spring runoff. Please note the warning on the map just past Illusion Creek.

Once you have conquered the difficulties, the river definitely slows down and widens out, and you can see the distant hill of the Yukon River Valley. To me, these always seem somewhat purple in color.

Forest fires have devastated the countryside during the last few years and the old village at the Yukon River confluence barely escaped destruction. It will take many years for things to return to normal.

A sure sign that you are close to the Yukon River is the overhead tramline running across the river. Canadian Government Water Survey uses the line to measure water volumes. Every sizeable stream in the Yukon is adorned with one of these.

As you approach Big Salmon Village, there are still a few potential trouble spots. There are a number of strong riffles coming off the rock bluffs in several curves close to the Yukon River. Stay to the inside of the curves.

Headless Creek—Mile 140 (224)

MAP 13 **PAGE 51**

H. S. Bostock of the Geological Survey of Canada, named this creek in the spring of 1934. Glaciation had diverted the headwaters of the creek into Lokken Creek, leaving this creek "headless."

Illusion Creek—Mile 142 (km 227)

MAP 13, 14 **PAGE 51, 52**

Again this was named by Bostock in 1934. Some of his survey party mistook this creek for Lokken Creek and missed their rendezvous with the main party.

Big Salmon Village—Mile 147 (km 235)

MAP 15 **PAGE 53**

This village seems to have been an encampment of sorts for the Yukon Natives as far back in memory as one can go. The first recorded trips into the Big Salmon all spoke of Indians encamped here and at the mouth of the North Fork upstream.

Gertie Tom, a Tutchone Native woman who was born at the village in 1927, recounts her life in and around the village in a 1987 publication called *My Country: Big Salmon River*. Gertie and her brothers and sisters were raised here. Other relatives came from such places as Tagish (on Marsh Lake), Little Salmon Village on the Yukon River and Ross River. There were a number of Native families living at the village from the 1920s to the 1940s. These included the families of Jim Shorty, Soo Bill, Harry Silverfox, George Peter and Billy Silverfox.

Gertie does not relate the history of the village prior to her birth, but we can assume it was a busy point on the river as it warranted a Northwest Mounted Police post. This was built on the upstream opposite bank to the village. One of the reasons for this isolation was that it was impossible for the large river steamers to land at the village due to the shoals at the mouth of the Salmon River. There also used to be quite a large graveyard attached to the village, but I understand that this was destroyed in the forest fires of 1994.

It is a great place to wind down from a Big Salmon River trip. As the village is also used by the Teslin and Yukon rivers' traffic, it is not unusual to see a number of canoes pulled up.

The rest of your trip to Carmacks is covered in my book *Exploring the Upper Yukon River—Part 1*. The second part of the Yukon River series takes you all the way to Dawson City.

The Big Salmon River. *Photo: Gus Karpes.*

4. Maps

The Big Salmon River

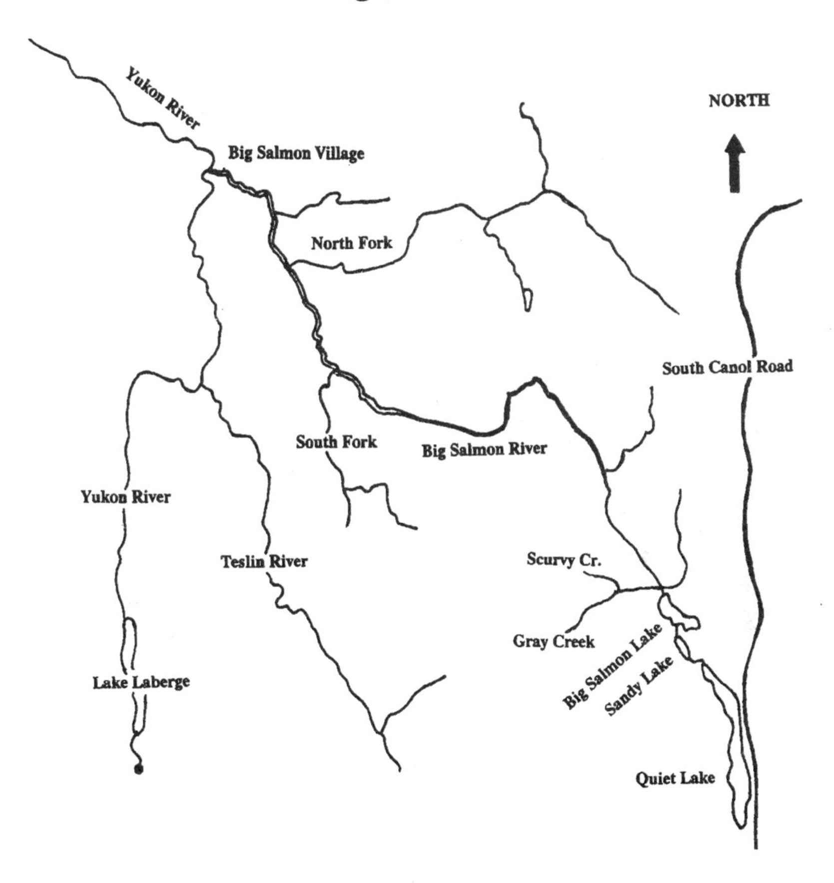

Contour lines in 500' increments are given throughout the map pages. This should give you an idea of the terrain immediately bordering the river.

mile/km — The distances shown indicate the mileage traveled from the put-in at Quiet Lake.

X ** — Potential campsites are marked with an X.
** Indicates that the space available is suitable for a group.

A sandbar or beach borders the river.
(At the higher water levels this may be under water.)

Note the North direction shown on each page. This is true north. Average declination in this country is from 33 to 34 degrees east. We recommend that you also bring along topographical maps 105F–Quiet Lake and 105E–Laberge.

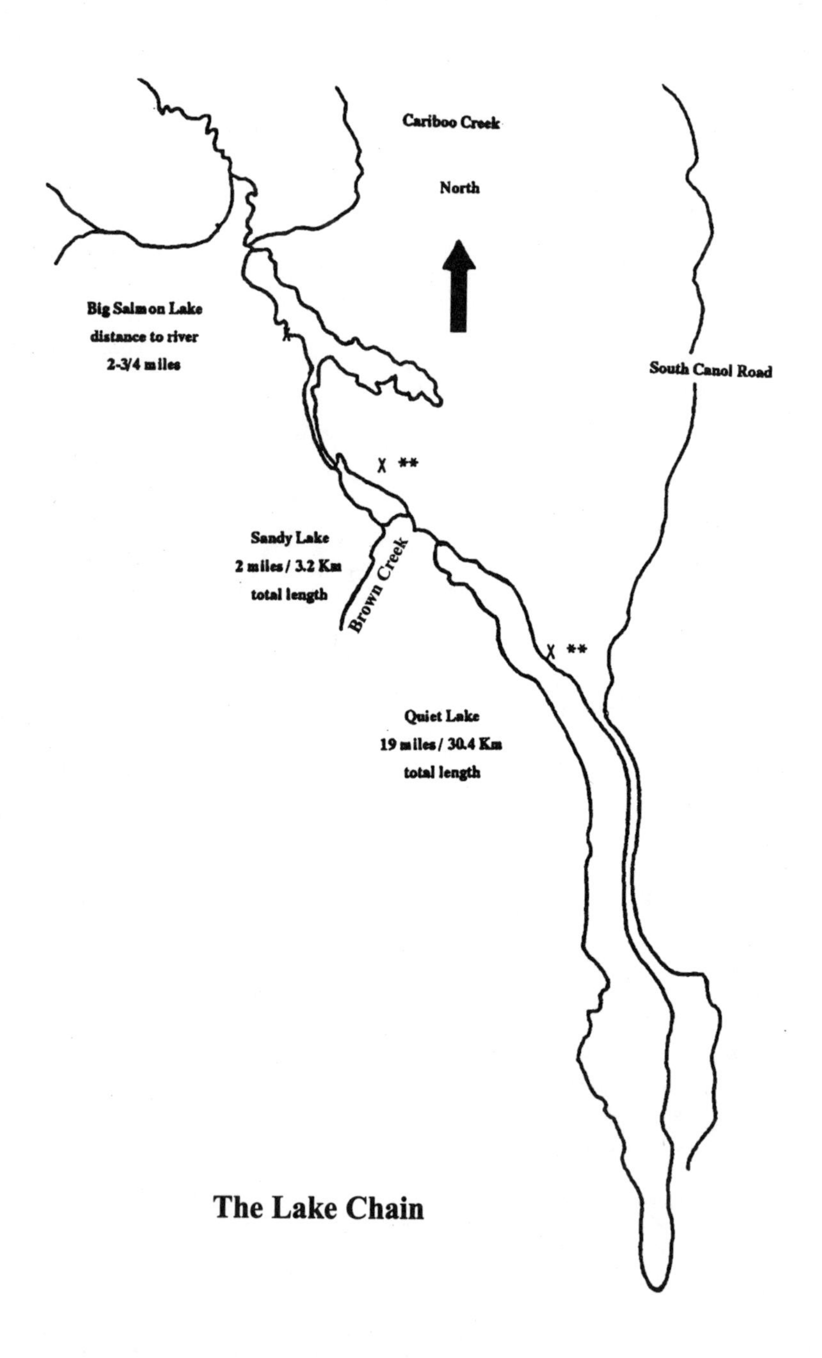
Cariboo Creek
North
Big Salmon Lake
distance to river
2-3/4 miles
South Canol Road
Sandy Lake
2 miles / 3.2 Km
total length
Brown Creek
Quiet Lake
19 miles / 30.4 Km
total length
The Lake Chain

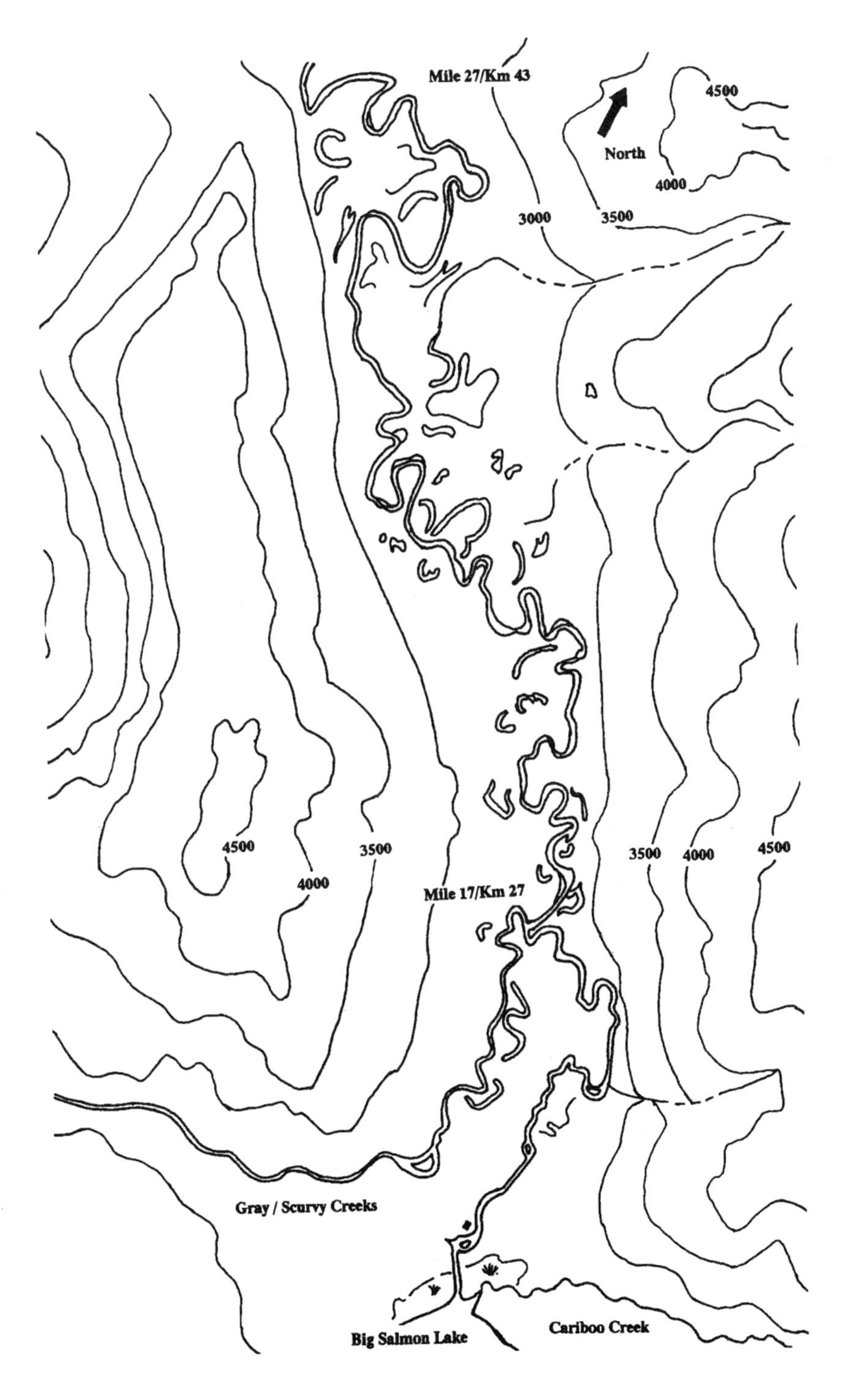

Mile 27/Km 43
North
4500
4000
3000
3500
4500
3500
4000
Mile 17/Km 27
3500
4000
4500
Gray / Scurvy Creeks
Big Salmon Lake
Cariboo Creek

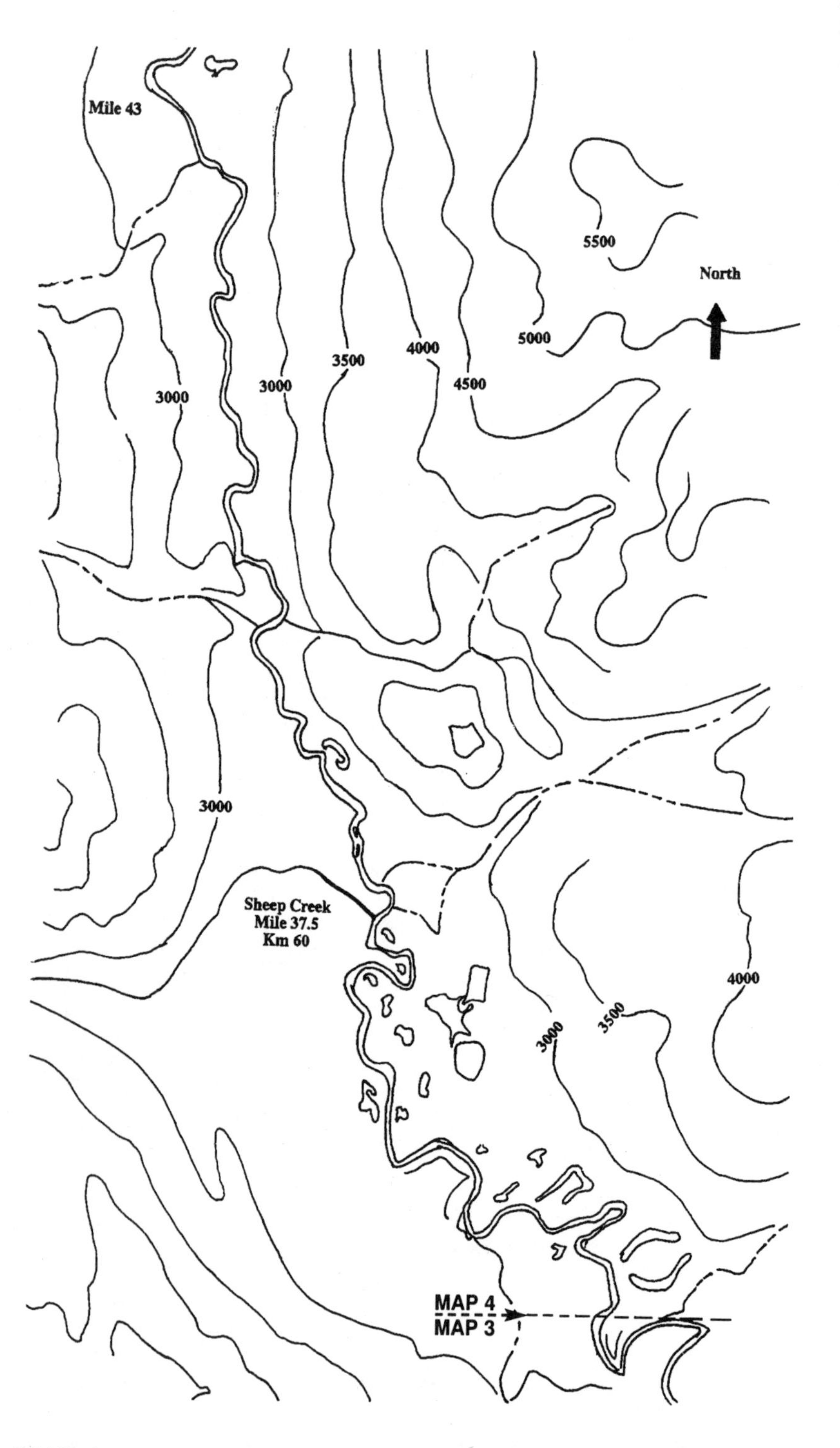
Mile 43
5500
North
5000
4000
3500
4500
3000
3000
3000
Sheep Creek
Mile 37.5
Km 60
4000
3000
3500
MAP 4
MAP 3

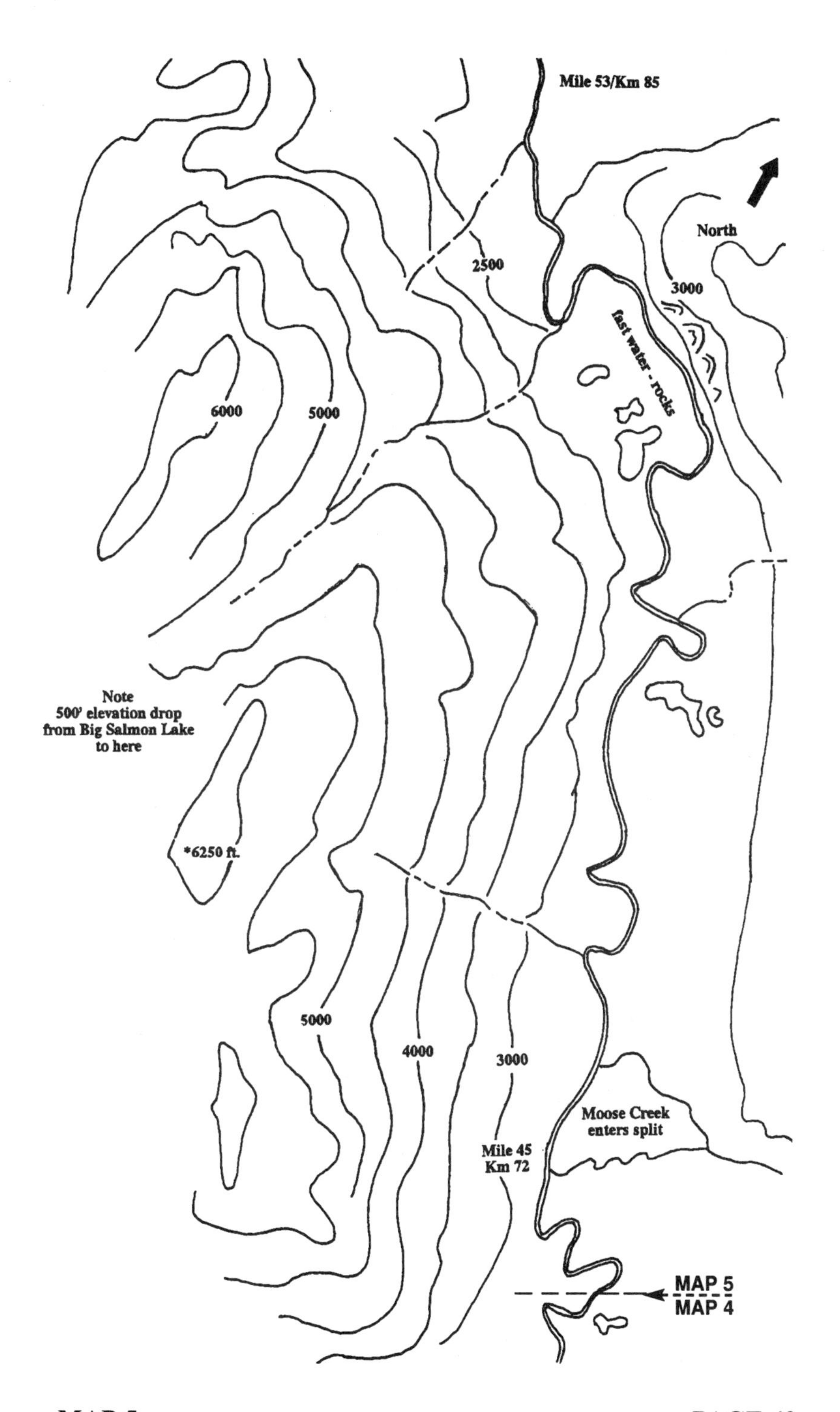
Mile 53/Km 85
North
2500
3000
fast water - rocks
6000
5000
Note
500' elevation drop
from Big Salmon Lake
to here
*6250 ft.
5000
4000
3000
Moose Creek
enters split
Mile 45
Km 72
MAP 5
MAP 4

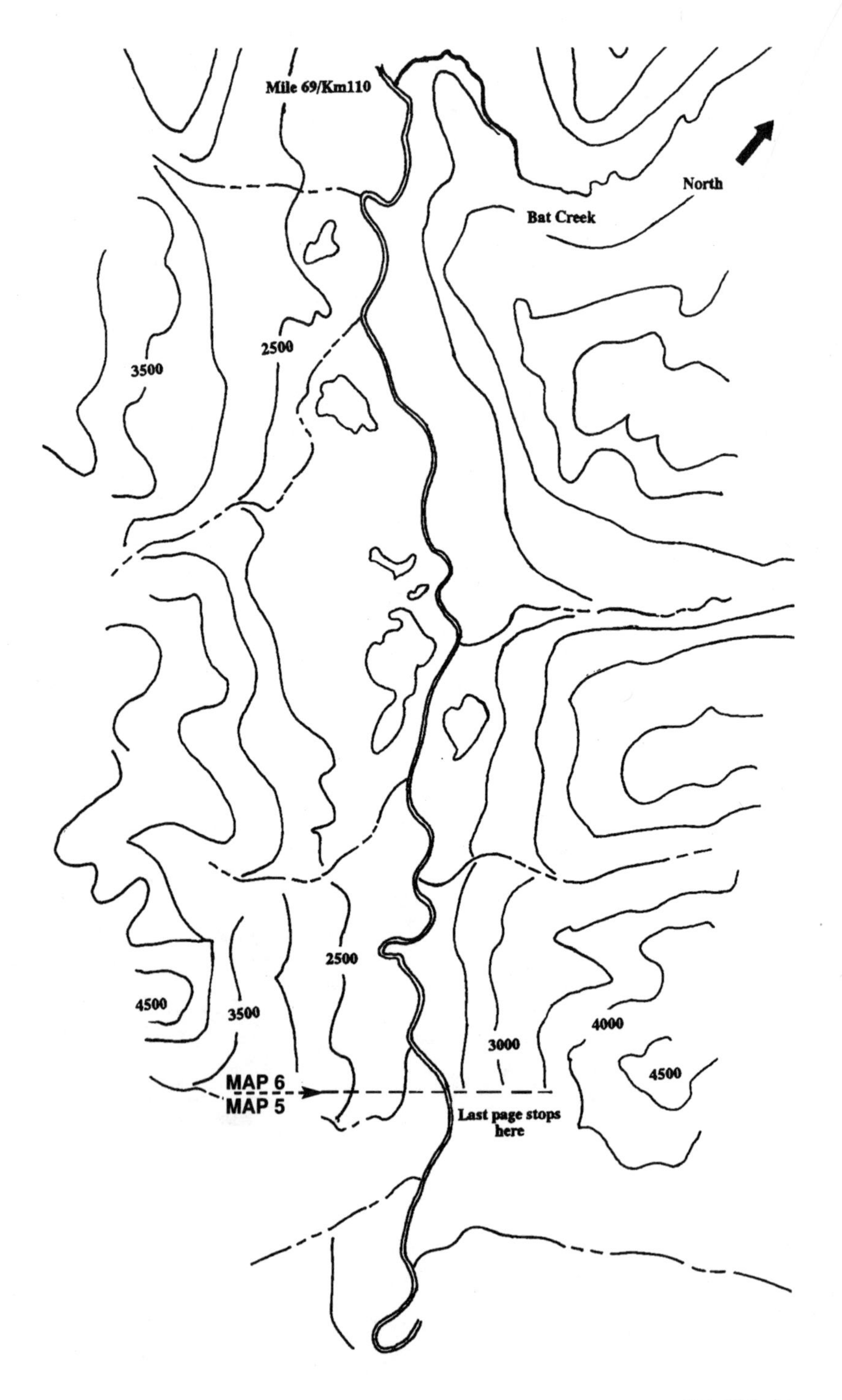
Mile 69/Km110
North
Bat Creek
2500
3500
2500
4500
3500
4000
3000
4500
MAP 6
MAP 5
Last page stops here

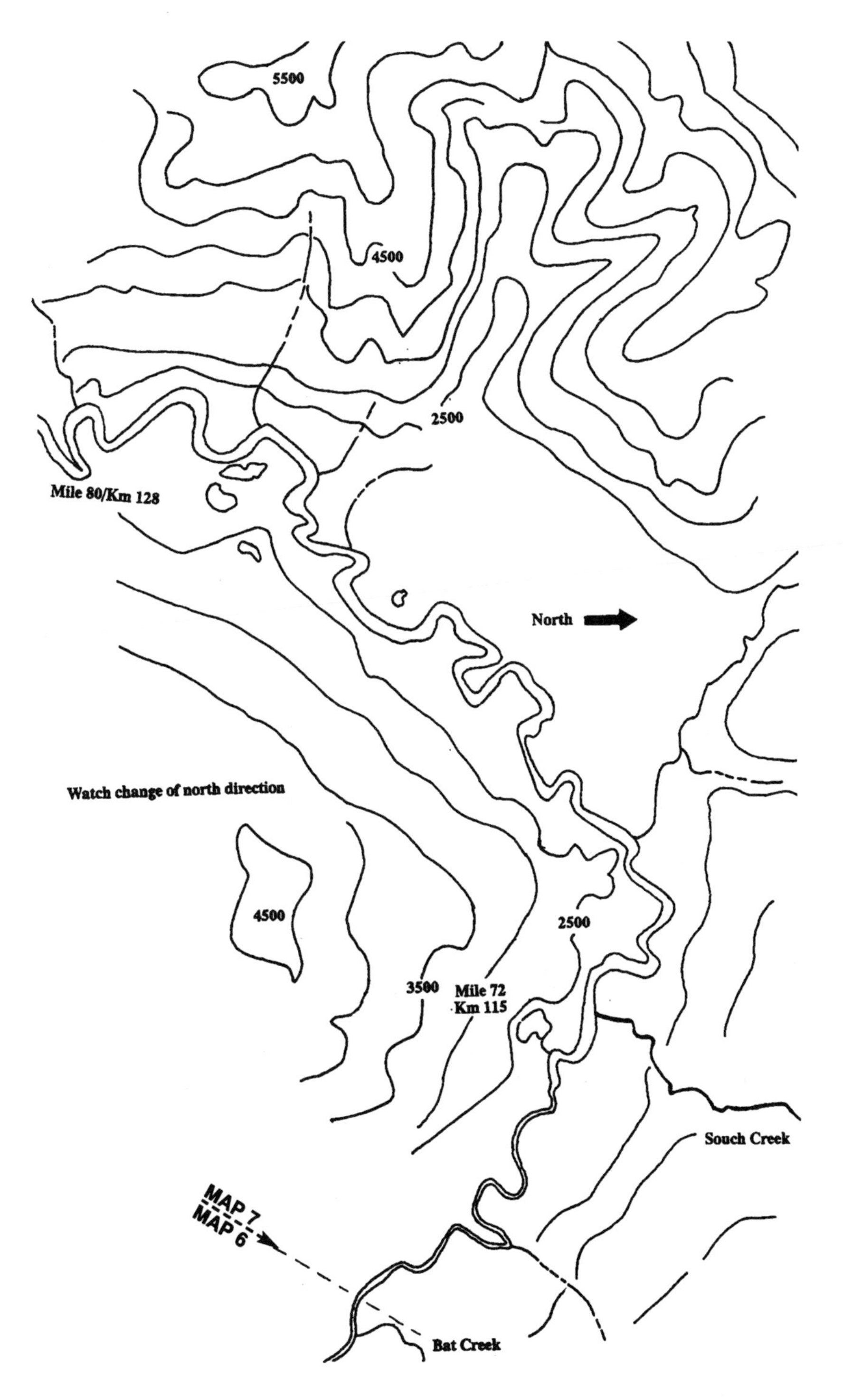
5500
4500
2500
Mile 80/Km 128
North
Watch change of north direction
4500
2500
3500
Mile 72
Km 115
Souch Creek
MAP 7
MAP 6
Bat Creek

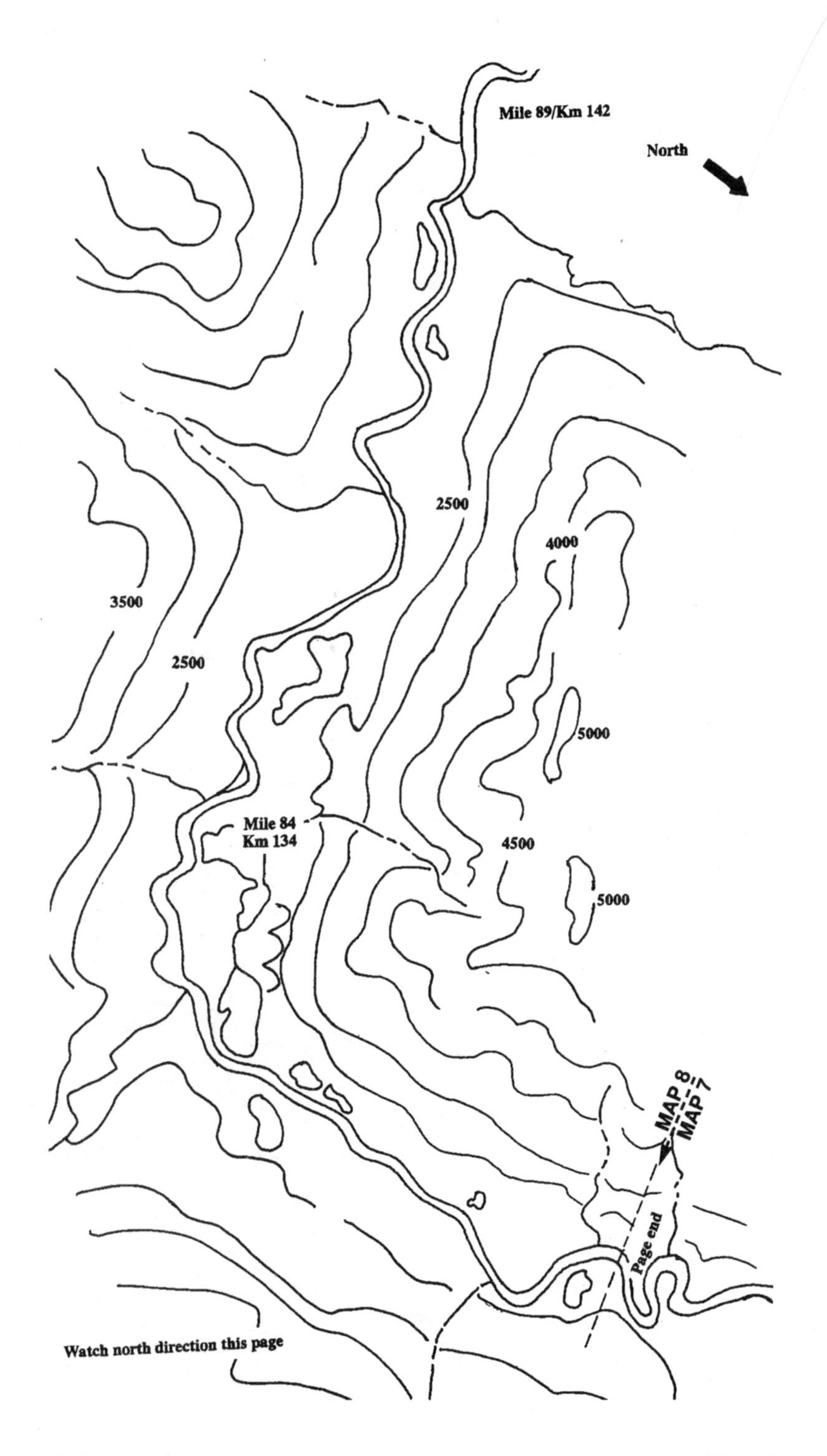
Mile 89/Km 142
North
2500
4000
3500
2500
5000
Mile 84
Km 134
4500
5000
MAP 8
MAP 7
Page end
Watch north direction this page

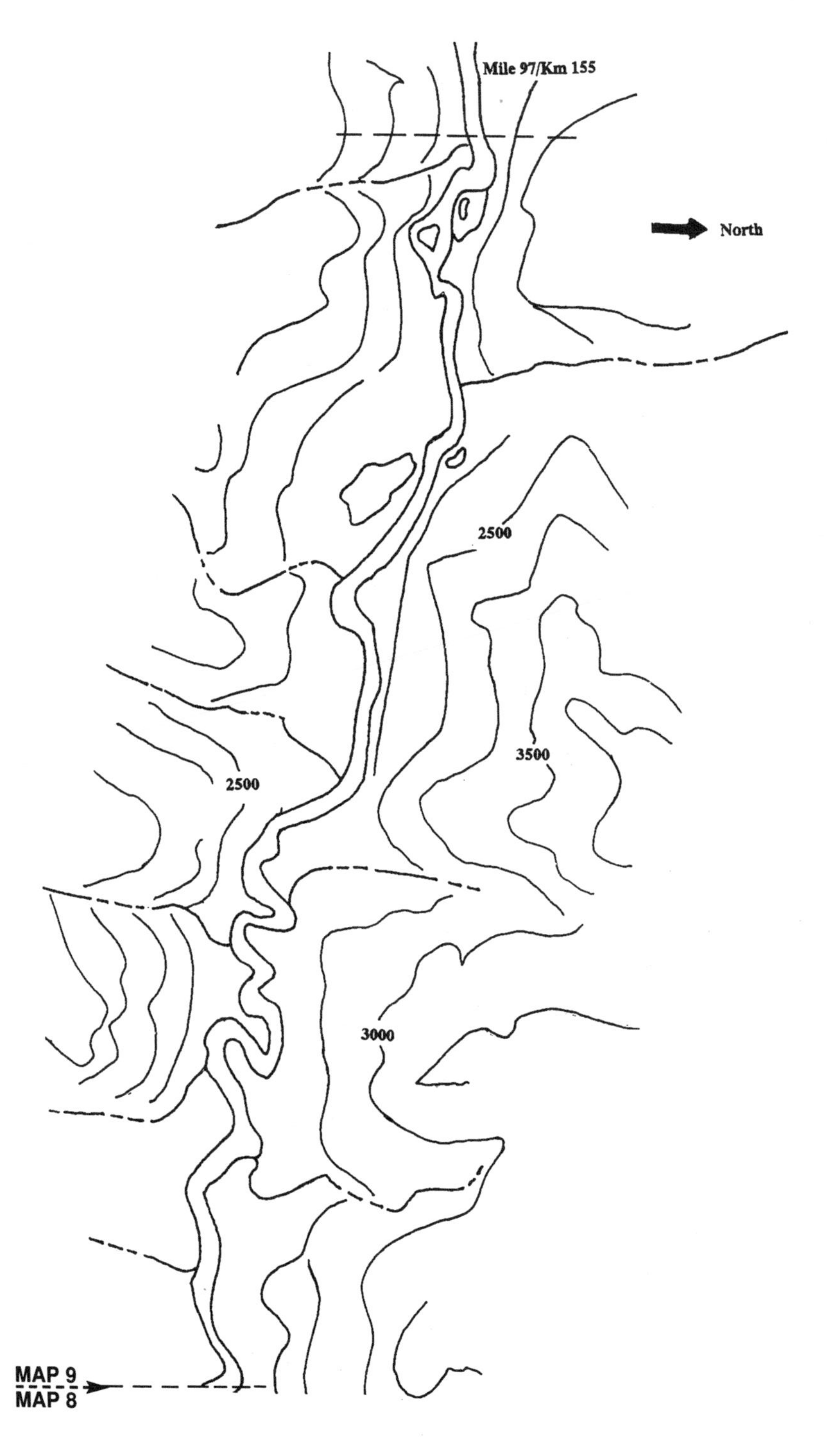
Mile 97/Km 155
North
2500
3500
2500
3000
MAP 9
MAP 8

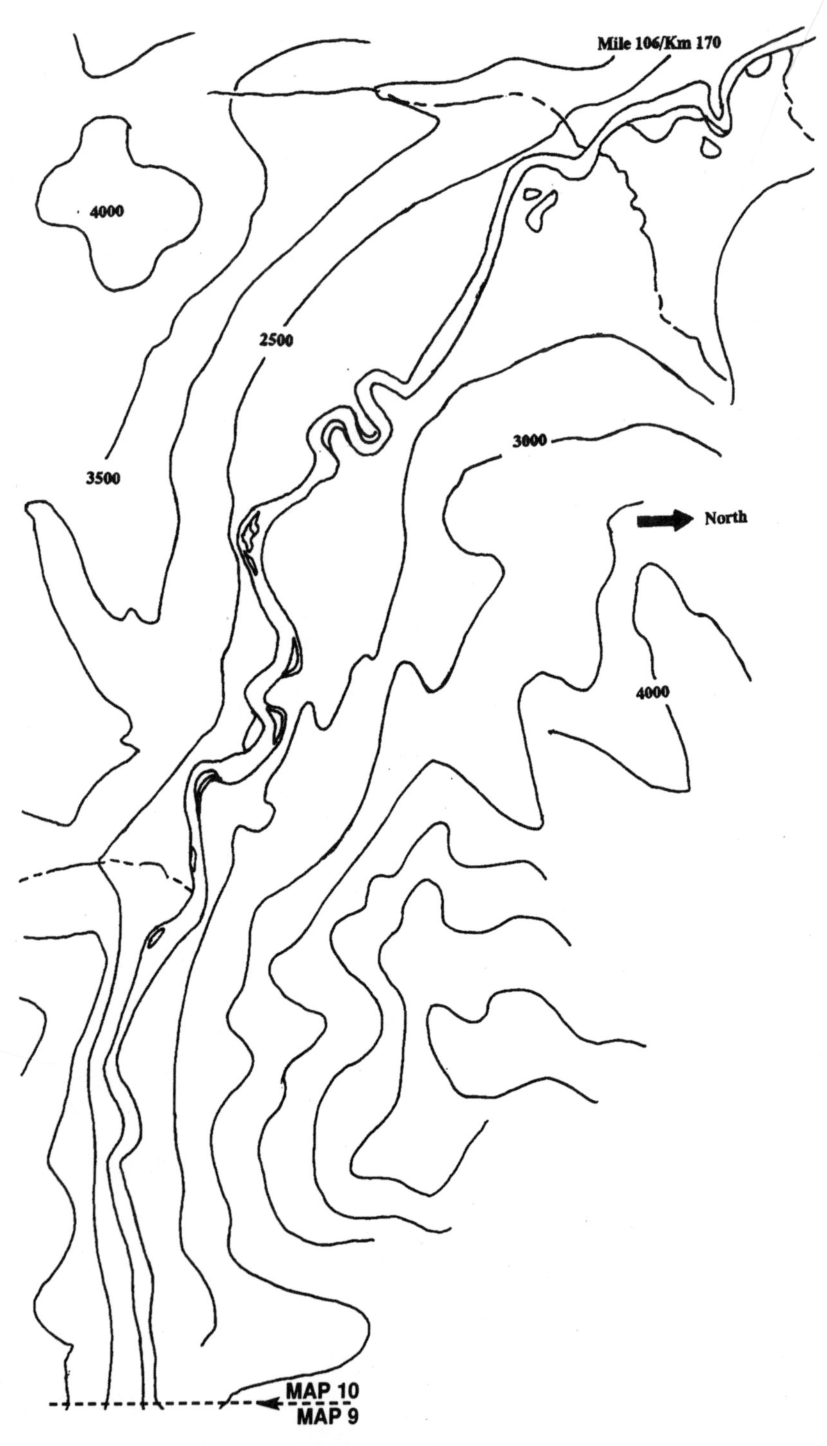
Mile 106/Km 170
4000
2500
3500
3000
North
4000
MAP 10
MAP 9

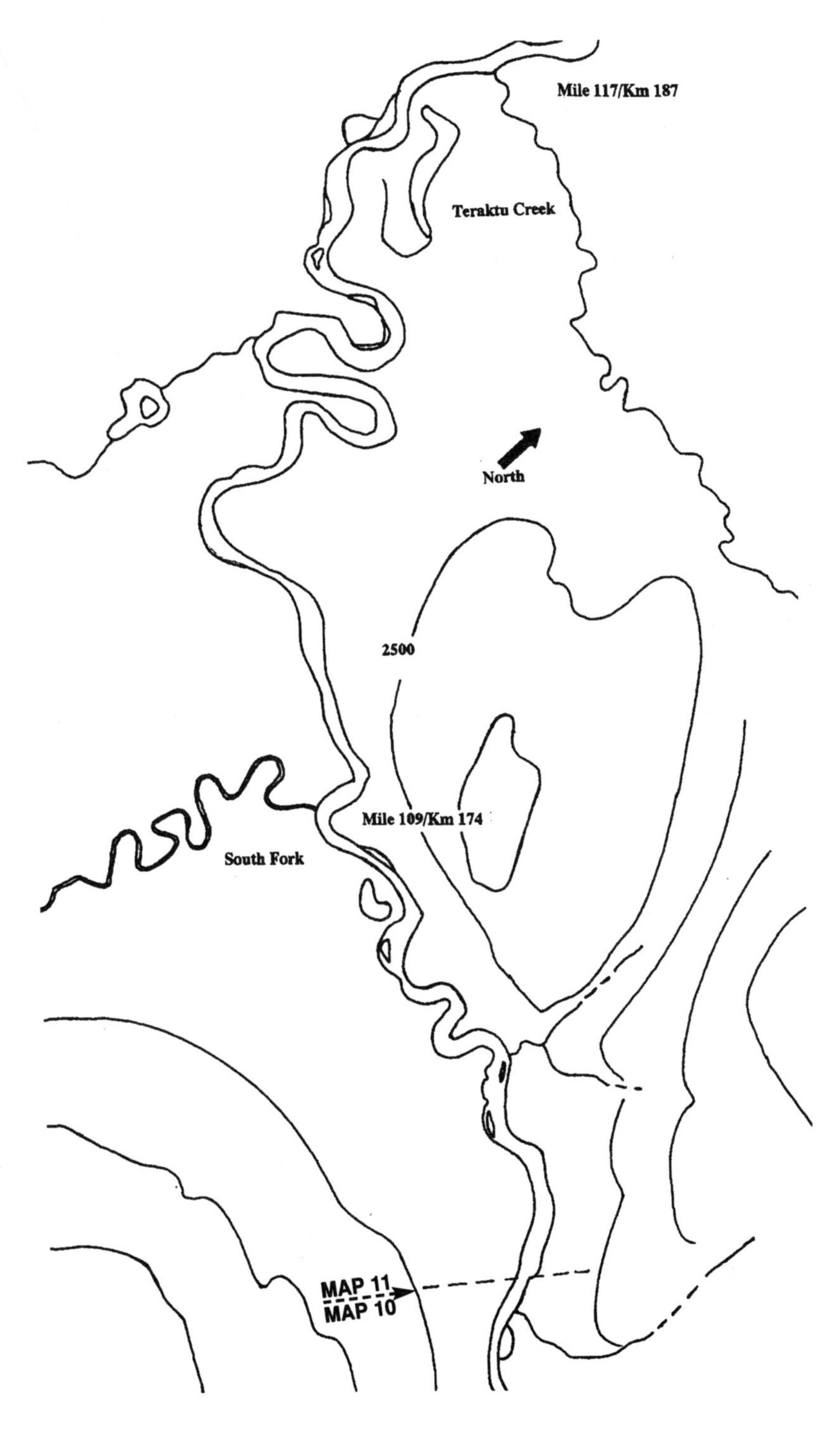
Mile 117/Km 187
Teraktu Creek
North
2500
Mile 109/Km 174
South Fork
MAP 11
MAP 10

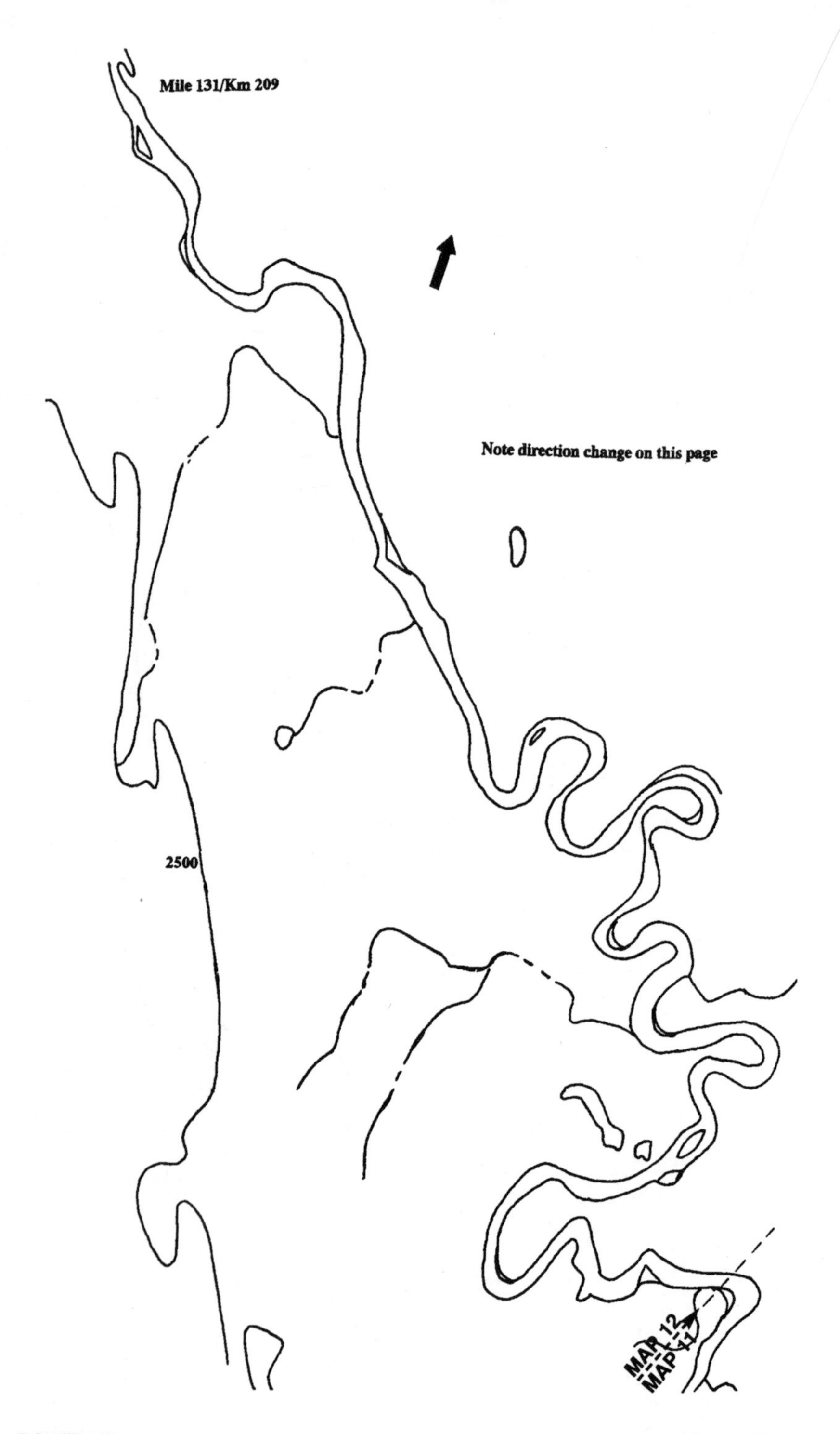
Mile 131/Km 209
Note direction change on this page
2500
MAP 12
MAP 11

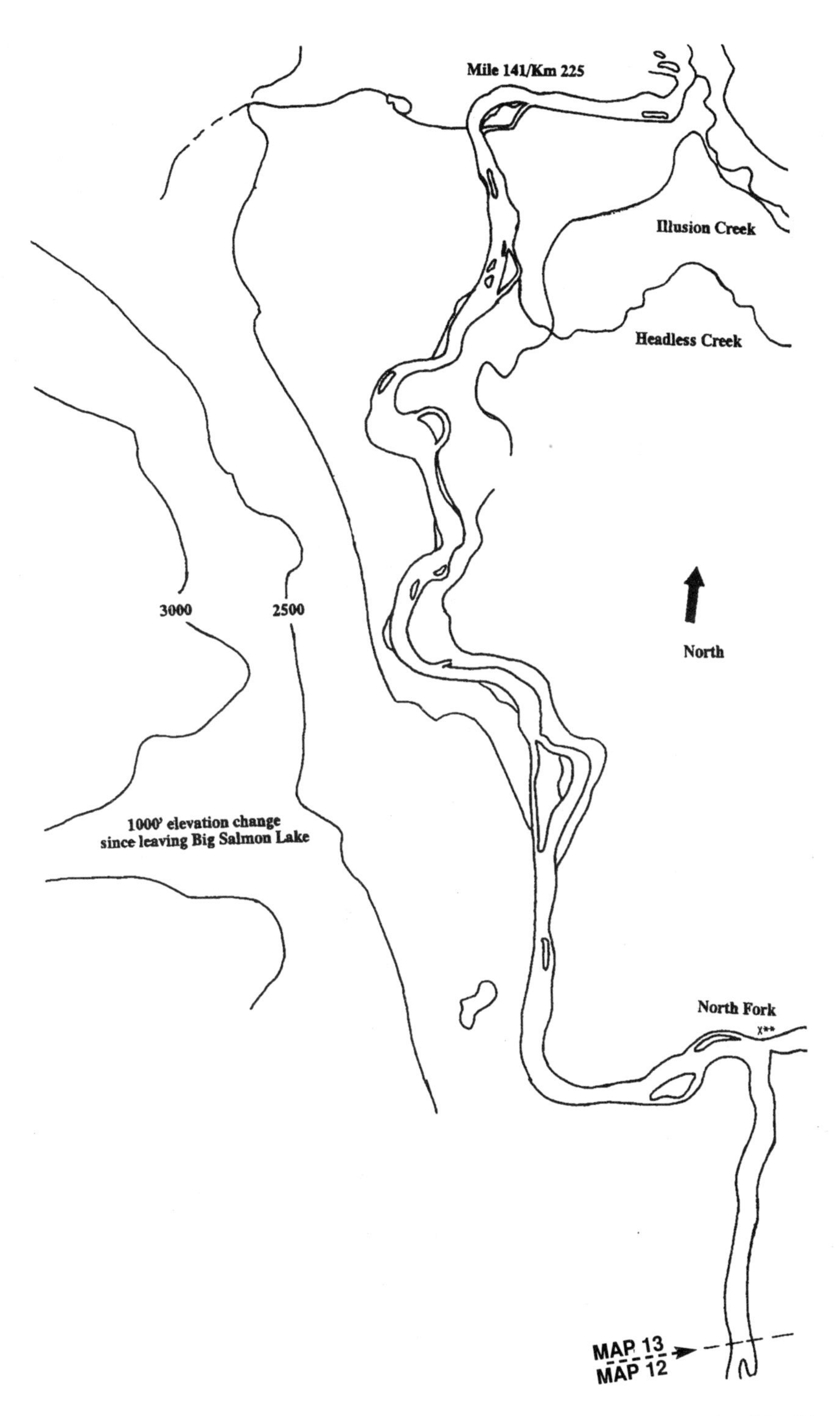
Mile 141/Km 225
Illusion Creek
Headless Creek
North
3000
2500
1000' elevation change
since leaving Big Salmon Lake
North Fork
MAP 13
MAP 12

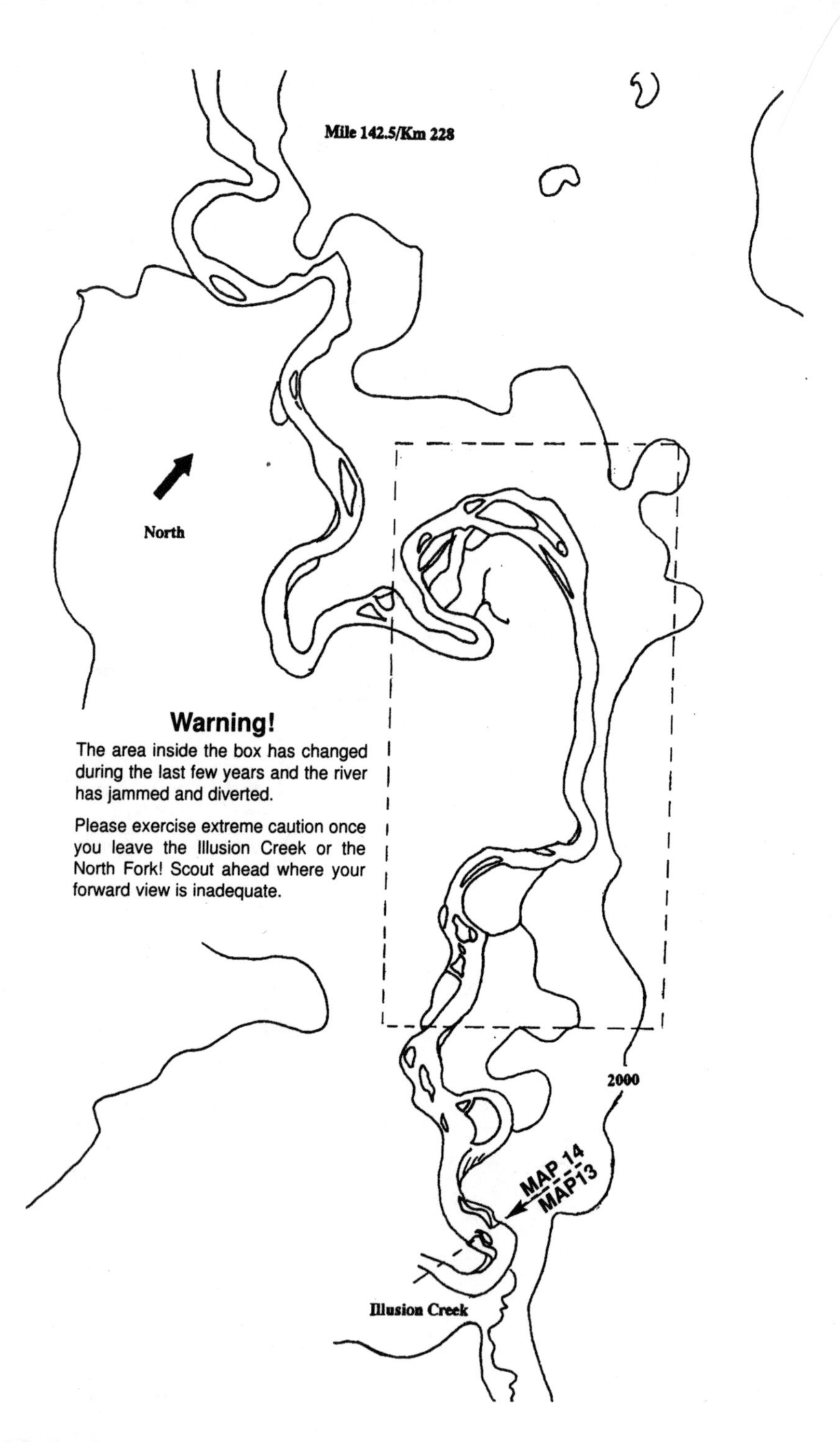
Mile 142.5/Km 228
North
Warning!
The area inside the box has changed during the last few years and the river has jammed and diverted.
Please exercise extreme caution once you leave the Illusion Creek or the North Fork! Scout ahead where your forward view is inadequate.
2000
MAP 14
MAP 13
Illusion Creek

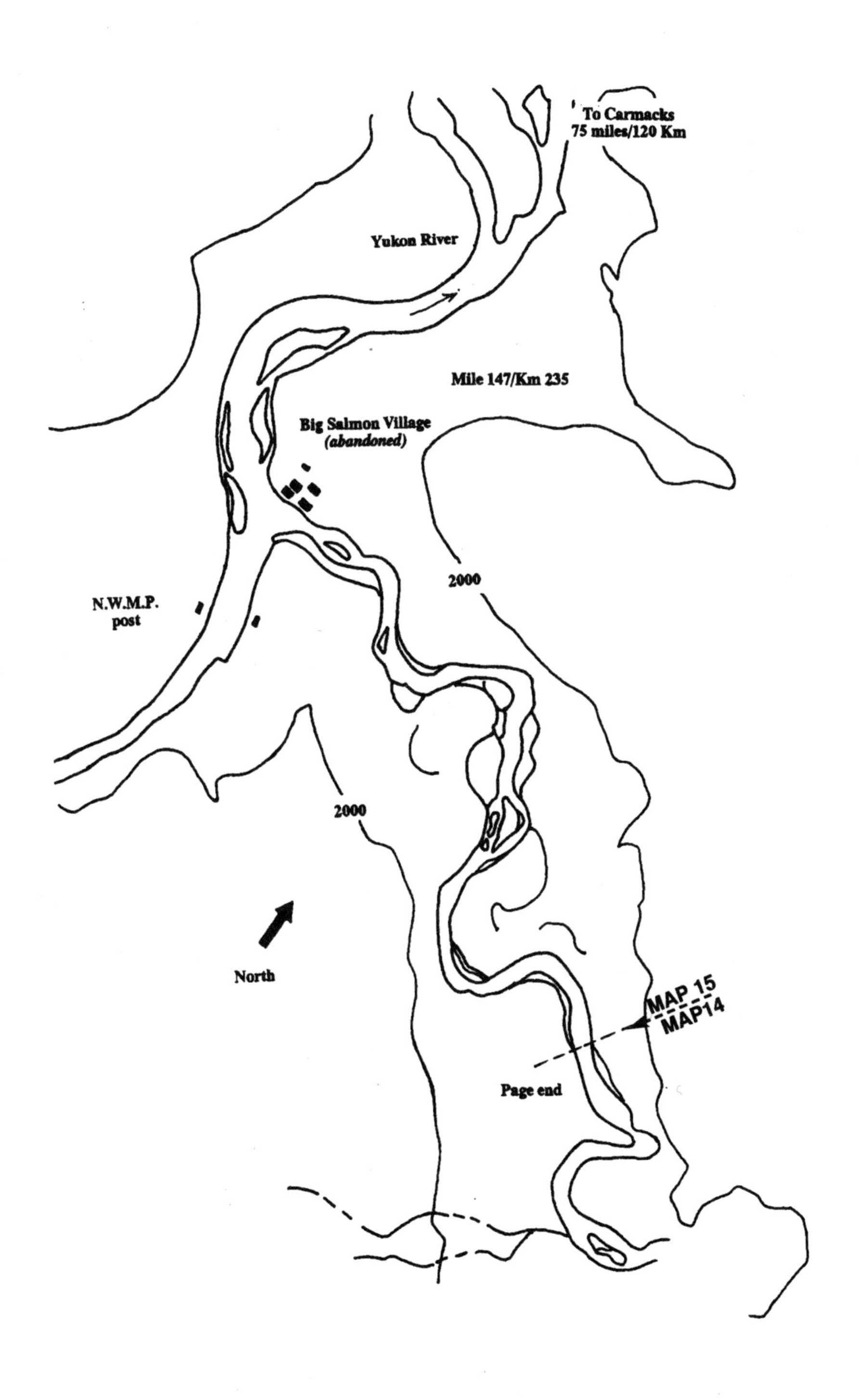
To Carmacks
75 miles/120 Km
Yukon River
Mile 147/Km 235
Big Salmon Village
(abandoned)
2000
N.W.M.P.
post
2000
North
MAP 15
MAP14
Page end

The Big Salmon River. *Photo: Scott McDougall.*

5

The Catch

The Arctic Grayling

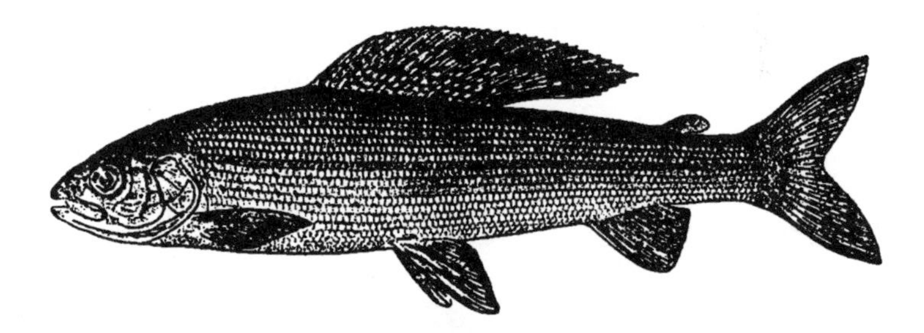

This is probably the most common fish encountered in the Salmon River watershed. It is abundant in the connecter streams between the lakes and also in the main river. It is a strikingly colored fish. The dorsal fin, its main identifying feature, is a dark purple to blue-black with a narrow mauve edge. In the males, the large dorsal fin can reach beyond the adipose fin; it is shorter for the females. In the males, the dorsal fin is highest at the back; in females the fin is higher at the front.

Grayling spawn in the smaller streams running into the main river during the period in which the ice is first breaking up. During this break up, the adults migrate from ice-covered lakes and larger rivers to the small gravel- or rock-bottomed tributaries. Where there are no suitable small streams for spawning, migration takes place in the rocky parts of the main rivers.

Maximum age of the fish is from eleven to twelve years. It reaches spawning maturity in about four years, although this may be extended to as much as six or eight years in the colder northern lakes and rivers. Maximum known size in Canada and the present angling record is 29 7/8 inches long and five pounds fifteen ounces in weight. The fish was caught in the Katseyedie River in the Northwest Territories in 1967.

The average size of the fish in the Big Salmon River system is between twelve and eighteen inches with an average weight of one pound. They are larger in the lakes where they are not subjected to the fast flow of the main river, giving them a chance to retain some of their meat as they don't burn up as much energy when swimming.

A light to ultralight fishing rod and reel with a four-pound test line is more than adequate to handle the fish. In a pinch, a small hook baited with a bumble bee or like insect will do the trick. The grayling normally feeds on all kinds of insects.

Do not keep more fish than you intend to eat. Take the barbs off the hook. Do not handle the fish if you intend to release it. If you do handle it, chances are it will die.

If frying the fish, scale it first. If baking it in foil, there is no need to scale it as the skin will fold back easily after cooking, taking the scales with it.

Lake Trout

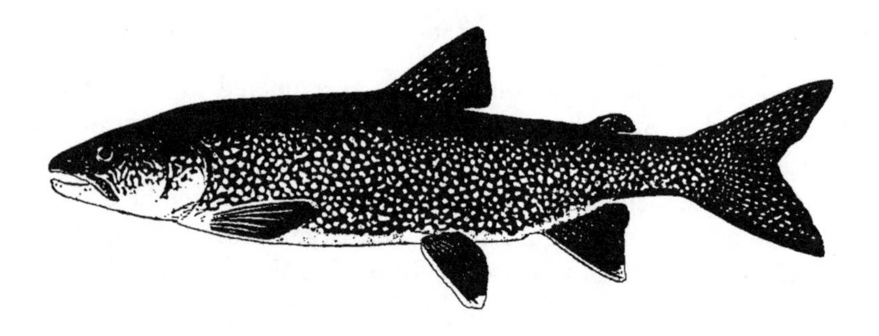

These fish are mostly found in the upper lakes at the beginning of the trip, although the mouths of side streams coming into the main Salmon River may also yield this delectable food. The exact coloration of the fish can change from lake to lake depending on what the fish has been eating. Overall coloration is light spots on a darker background with light-colored underbelly. The whole body, including the head, dorsal and adipose fins are covered with innumerable light-colored spots. The flesh may be white, pink, orange or orange-red. An average lake trout in this system is between fifteen and twenty inches and will weigh between two and four pounds.

Again, light spinning tackle will do the trick. The trout seems to prefer the "spoon" type lure as opposed to the more active spinners preferred by the grayling.

Cooking is the same as the grayling, although you may have to cut up the trout to make it fit the pan.

Chinook Salmon

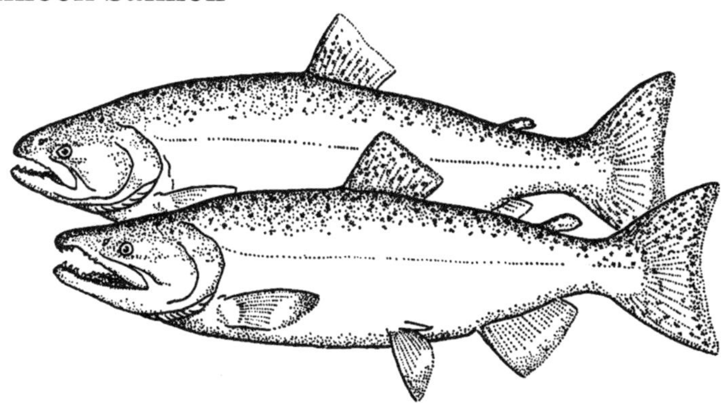

When you travel the Big Salmon River in August or September, you cannot help but notice these fish in the system. Although when this large Pacific salmon enters the fresh water of the Yukon River they are green in color, the spawning fish that you are likely to encounter are an olive-brown to purply color with the males slightly darker than the females. The spawning male also develops a pronounced hooked snout and protruding teeth. This is why it is sometimes referred to as the "dog salmon."

These fish have traveled almost 2,000 miles to get here. During the journey they have had to run a gauntlet of nets and fish wheels in addition to all of the natural hazards. The fish wheel is particularly deadly. It is a wheel with paddled spokes mounted on a raft. This is set into the river and the current turns the wheel in the manner of a water wheel of the past. Wire baskets are attached to the paddles. These literally scoop the fish out of the water and drop them in a holding tank strategically placed on the raft to catch the salmon falling out of the baskets on the way down. It requires little attendance, other than an occasional emptying of the fish tank, and goes twenty-four hours per day.

Although it is legal for you to catch spawning salmon, I feel that they should be left alone to procreate in peace without further interference from us.

Cleaning Fresh Fish

Many times, while guiding people on river trips, I incorrectly assume that everyone knows how to clean a fish. This is not so. Many have never been that close to a primary food source. Most are familiar with fish served cooked in a restaurant or the fresh variety sold in a supermarket—that is cleaned, deboned and packaged in cellophane ready for the cook. Here are some basic cleaning instructions.

You should have a sharp knife and some sort of cutting board. For a cutting board, use a piece of driftwood. If nothing else is handy, use the blade of your paddle, making sure you don't cut into it. A rock is a little hard on the knife blade.

Scale the fish. If your camp knife is equipped with a fish scaler all the better, if not, use the knife on edge, the sharp side down and "scrape" tail to head direction. Do not cut into the fish. Remove all of the scales and again wash the fish.

To gut the fish, hold it upside down in one hand or lay it down on the cutting board if you are having trouble holding on. Slit the fish from the vent forward to the gills. Do not cut too deep and try to keep the knife just under the skin. Open up the fish and clean out the cavity.

If you are not into eating the head, cut it off just behind the gills. Wash the fish thoroughly. It is now ready to cook.

I do not remove the fins, nor do I try to fillet a small fish. It is much easier to do this after the fish has cooked.

If frying the fish, cook it on low heat otherwise it will curl up badly in the pan. Cover the pan while cooking the first side. Turn only once and leave uncovered for the remainder of the cooking. Short of burning it, it is almost impossible to overcook fish.

If baking the fish in foil, double wrap the fish and seal it tight. It will poach in its own juices. Again, use medium to low heat.

Good luck!

Photo: Scott McDougall.

Deaf Charlie

I met Charlie briefly a number of years ago. Obviously he made an impression on me as I have no problem recalling the short encounter. Memories of Charlie are synonymous with the recall of Arctic grayling gently frying in the pan over an early morning campfire. This day-dream frequently flits through my mind during the long, cold months of a northern winter.

Charlie was a short, somewhat smallish person. His slightly stooped frame bespoke of a comfortable familiarity with life. His age-wrinkled, tanned features were topped with a battered, one-size-too-large canvas fedora. His slightly cauliflowered ears, in addition to anchoring the hat, had as their most distinguishing feature bright, shiny, silver hearing-aids in the shape of a seashell protruding from them.

In my occupation as an outdoor guide, I meet a great number of people during a summer. My personal trick for remembering them is to pick out an easily observed, distinguishing feature or trait about the person, which would trigger my memory when necessary.

Let's see, there has been Lame Bob, Henpecked George, Tall John and Joseph the German to name just a few. Charlie obviously, because of the silver accouterments, became Deaf Charlie.

At our first meeting at the airport, it became immediately obvious that Charlie used the silver horns both ways.

"How was the flight?" brought no response. I tried again.

"Nice day, eh!?" (It is expected of us Canadians to use a generous amount of ehs in our conversation with foreign tourists.) Again, no notable acknowledgment or response.

I was about to upgrade Charlie's standing to deaf mute, when he casually reached into both ears and with a slight turn and electronic squeal, there he was.

It was obvious that he had the habit of tuning himself out when the world around him became a little too noisy and he needed a little peace and quiet. I am sure many of us can relate, but of course we don't have that ability or option to isolate ourselves at will.

The very first morning of our ten-day canoe trip, Charlie established himself as a fisherman extraordinaire. Up early, he nonchalantly picked a fishing rod out of the community gear and without so much as identifying the hardware on the end of the line, proceeded to catch a sufficient number of Arctic grayling to satisfy everyone for breakfast.

"Ha!" I thought to myself, "Beginner's luck. We'll see how it holds out once we're downriver a ways."

Much to my chagrin, things stayed the same. Almost before I got up each morning, Charlie had breakfast caught, cleaned, scaled and awaiting the frying pan.

Now, freshly caught grayling for breakfast is a rare and wonderful experience that everybody should have the opportunity of trying at some time of their life. You can understand however, that as the trip progressed, one by one the rest of the crew opted out and settled for the more mundane items on the breakfast menu. Charlie's uncanny ability to gauge his customers continued to astound me. The number of fish he brought in for breakfast, automatically seemed to adjust itself. He, of course, would eat his pan-fried fish each breakfast with the same lip-smacking relish he exhibited each and every morning of the trip.

Charlie was quite the character. He did not personally own a fishing rod and for that matter, told me that he was not in the habit of eating fish at home, not even the store-bought, cellophane-wrapped variety. On a wilderness outing he reverted back to the primeval instincts of the provider. To me anyway, he seemed to have a communion with the river gods who, without fail, opened up their finny larder for Charlie each morning.

I will always remember him as Deaf Charlie. In retrospect, he is deserving of a more exciting reference such as Never-Miss Charlie, Lucky Charlie or Grayling Charlie. Ah well! I might then lose my most distinguishing reference to Charlie and a treasured memory in the process.

References

Berton, Pierre. *Klondike—The Last Great Gold Rush*, McClelland and Stewart Limited, Toronto, 1972.

Bostock, H. S. Compiled and Annotated. *Yukon Territory—Selected Field Reports of the Geological Survey of Canada 1898 to 1933*, Memoir 284 Geological Survey of Canada, 1957.

Coutts, R. *Yukon Places & Names*, Gray's Publishing Limited, Sidney, British Columbia, 1980.

Dawson, George M., DS, FGS *Report on an Exploration in the Yukon District, N.W.T. and Adjacent Northern Portion of British Columbia—1887*, Reprinted with permission from Geological and Natural History Survey of Canada, 1987. (Yukon Historical and Museums Association.)

Hamilton, W. R. *The Yukon Story*, Mitchell Press Limited, Vancouver, Canada, 1964.

Harris, A. C. *Alaska and the Klondike Gold Fields*, "Practical Instructions for Fortune Seekers", etc. 1897, Facsimile edition published by Coles Publishing Company, 1972.

Johnson, James Albert. *Carmack of the Klondike*, Epicenter Press and Horsdal & Schubart, 1990.

Schwatka, Frederick. *A Summer in Alaska*, J. W. Henry, St. Louis, MO, 1894.

Scott, W. B. and E. J. Crossman. *Freshwater Fishes of Canada*, Bulletin 184—Fisheries Research Board of Canada, 1973.

Tom, Gertie. *My Country: Big Salmon River*, Yukon Native Language Centre, 1987.

Wilson, Clifford. *Campbell of the Yukon*, Macmillan of Canada, Toronto, 1970.

Wright, Allen A. *Prelude to Bonanza*, Gray's Publishing Ltd., Sidney, British Columbia, 1976.

Index

The Yukon series by Gus Karpes

Exploring the Big Salmon River
Quiet Lake to the Yukon River
5½ x 8½, SC, 64 pp.
ISBN 0-88839-422-5

Exploring the Upper Yukon River
Whitehorse to Carmacks
5½ x 8½, SC, 134 pp.
ISBN 1-896407-04-8

The Teslin River
Johnson's Crossing to Hootalinqua, Yukon
5½ x 8½, SC, 102 pp.
ISBN 1-896407-00-5

Exploring the Upper Yukon River
Carmacks to Dawson City
5½ x 8½, SC, 112 pp.
ISBN 0-88839-421-7

The Nisutlin River
Mile 40 South Canol Road to Teslin, Yukon
5½ x 8½, SC, 50 pp.
ISBN 0-88839-422-X

OTHER HANCOCK HOUSE NATURE TITLES

Alpine Wildflowers
J. E. Underhill
ISBN 0-88839-975-8

Backroads Explorer:
Similkameen & S. Okanagan
Murphy Shewchuck
ISBN 0-88839-205-2

Clancy & Tidepool Friends
Carol Batdorf
ISBN 0-88839-336-9

Coast Lowland Wildflowers
J. E. Underhill
ISBN 0-88839-973-1

Eastern Mushrooms
E. Barrie Kavasch
ISBN 0-88839-091-2

Eastern Rocks & Minerals
James W. Grandy
ISBN 0-88839-105-6

Edible Seashore
Rick M. Harbo
ISBN 0-88839-199-4

Exploring the Outdoors: SW BC
Eberts & Grass
ISBN 0-88839-989-8

Introducing E. Wildflowers
E. Barrie Kavasch
ISBN 0-88839-092-0

Northeastern Wild Edibles
E. Barrie Kavasch
ISBN 0-88839-090-4

Orchids of NA
Dr. William Petrie
ISBN 0-88839-089-0

Pacific Wilderness
Hancock, Hancock & Stirling
ISBN 0-919654-08-8

Rafting in BC
VanDine & Fandrich
ISBN 0-88839-985-5

Roadside Wildflowers NW
J. E. Underhill
ISBN 0-88839-108-0

Rocks & Minerals NW
Leaming & Leaming
ISBN 0-88839-053-X

Sagebrush Wildflowers
J. E. Underhill
ISBN 0-88839-171-4

Seashells of the NE
Gordon & Weeks
ISBN 0-88839-808-7

Tidepool & Reef
Rick M. Harbo
ISBN 0-88839-039-4

Trees of the West
Mabel Crittenden
ISBN 0-88839-269-9

Upland Field & Forest Wildflowers
J. E. Underhill
ISBN 0-88839-174-9

Western Mushrooms
J. E. Underhill
ISBN 0-88839-031-9

Western Seashore
Rick M. Harbo
ISBN 0-88839-201-X

Wild Berries of the NW
J. E. Underhill
ISBN 0-88839-027-0

Wildflowers of the West
Crittenden & Telfer
ISBN 0-88839-270-2

Wild Harvest
Terry Domico
ISBN 0-88839-022-X

Wildlife of the Rockies
Hancock & Hall
ISBN 0-919654-33-9